C̶R̶O̶DUCTION SCIENCE IN HORTICULTURE series

Se̶ri̶es̶ e̶di̶tors: Jeff Atherton, Senior Lecturer in Horticulture, U̶ni̶ve̶rsi̶ty̶ o̶f̶ Nottingham, and Alun Rees, Horticultural C̶o̶ns̶ul̶ta̶nt̶ and Editor, *Journal of Horticultural Science*.

Th̶i̶s̶ s̶e̶ri̶e̶s̶ e̶xamines economically important horticultural crops selected from the m̶aj̶o̶r̶ s̶ys̶tems in temperate, subtropical and tropical climatic areas. Sy̶s̶te̶m̶s̶ o̶f̶ p̶ro̶du̶ct̶io̶n̶ range from open field and plantation sites to protected plastic an̶d̶ g̶la̶s̶s̶-̶h̶o̶u̶s̶e̶ growing rooms and laboratories. Emphasis is placed on the sc̶ie̶nt̶if̶ic̶ p̶ri̶nc̶ip̶le̶s̶ underlying crop production practices rather than on providing en̶cy̶cl̶o̶p̶ae̶di̶c̶ a̶critical acceptance. Scientific understanding provides the ke̶y̶ t̶o̶ t̶h̶e̶ c̶ho̶ic̶e̶ o̶f̶ practice and the solution of future problems.

Students and teachers at universities and colleges throughout the world involved in co̶ur̶s̶e̶s̶ i̶n̶ h̶o̶rt̶ic̶ul̶tu̶re̶, as well as in agriculture, plant science, food science and ap̶pl̶ie̶d̶ b̶io̶lo̶gy̶ a̶t̶ d̶e̶gree, diploma or certificate level will welcome this series as a su̶pp̶le̶m̶e̶nt̶ar̶y̶ t̶ex̶tb̶oo̶k̶ source of information. The books will also be invaluable to pr̶o̶fe̶ss̶io̶na̶l̶ h̶o̶rt̶ic̶ul̶tu̶ri̶sts, advisers and end-product users requiring an authoritative, but br̶oa̶d̶ t̶re̶at̶me̶nt̶ o̶f̶ p̶ro̶duction to particular crops or systems. Amateur gardeners wishing to̶ u̶nd̶er̶st̶an̶d̶ t̶he̶ s̶cientific basis of recommended practices will also find the series ve̶ry̶ u̶se̶fu̶l̶.

The authors are internationally renowned experts with extensive experience of̶ t̶he̶ir̶ s̶ub̶je̶ct̶s̶. E̶a̶ch̶ v̶o̶lume follows a common format covering all aspects of pr̶o̶du̶ct̶io̶n̶, f̶ro̶m̶ s̶tr̶u̶ct̶ure̶ and physiology and breeding, to propagation and planting, th̶ro̶ug̶h̶ h̶us̶ba̶nd̶ry̶ a̶nd̶ crop protection, to harvesting, handling and storage. Se̶le̶ct̶ed̶ r̶ef̶er̶en̶ce̶s̶ a̶re̶ included to direct the reader to further information on sp̶ec̶if̶ic̶ t̶o̶pi̶cs̶.

CUCURBITS

R.W. Robinson
Cornell University
Department of Horticultural Sciences
Geneva, NY 14456
USA

and

D.S. Decker-Walters
The Cucurbit Network
P.O. Box 560483
Miami, FL 33256
USA

CAB INTERNATIONAL

CABI Publishing is a division of CAB International

CABI Publishing
CAB International
Wallingford
Oxon OX10 8DE
UK

CABI Publishing
875 Massachusetts Avenue
7th Floor
Cambridge, MA 02139
USA

Tel: +44 (0)1491 832111
Fax: +44 (0)1491 833508
Email: cabi@cabi.org
Web site: www.cabi-publishing.org

Tel: +1 617 395 4056
Fax: +1 617 354 6875
Email: cabi-nao@cabi.org

A catalogue record for this book is available from the British Library, London, UK

ISBN-13: 978-0-085199-133-7
ISBN-10: 0-085199-133-5

First published 1997
Reprinted with corrections 1999
Reprinted 2004
Transferred to print on demand 2006

Printed and bound in the UK by Antony Rowe Limited, Eastbourne.

CONTENTS

PREFACE

Scientists attempt to make the world more understandable, enjoyable and usable. Horticulture, in that it teaches us how to control one of our food sources – the plant kingdom, is an important practical science with a long history. Early investigative research in any field strives to develop general theories for what are perceived as the common state of affairs. The practical benefit of this approach is that, for example, we do not need to test every plant to know that it requires water and a source of assimilates to grow. However, evolution has produced a heterogeneous plant kingdom, requiring more detailed studies of distinct parts of the plant world in order to maximize our knowledge and efficacious use of individual crops.

The Cucurbitaceae is one of the most genetically diverse groups of plants in the plant kingdom. As a family and as individual crops, cucurbits epitomize adaptive differentiation and evolutionary divergence. Not only may cultivars within a crop vary significantly in their characteristics, but the same cultivar grown in distinct areas can have different needs in response to disparate local growing conditions. Cultures and ethnic groups may have different cultivar preferences and horticultural practices, which also increase morphological diversity within a crop. Consequently, the most difficult sentences in this book to write were those beginning with 'Cucurbits are ...' or 'Cucurbit crops require ...' or any sentence which attempted to generalize about this multifarious group of plants.

Crop production science is shifting gears from the general to the specific. Uniformity in practice is giving way to a new respect for the value of localized farming systems and the use of genetically diverse resources. In research, we are now accounting for a greater number of the many variables present in any biological investigation.

This expansion into detailed study has created a burgeoning body of research on cucurbits. We could not begin to cover all the pertinent information in one text. Instead, we hope that this book will give the reader a general awareness of cucurbit crop production, an understanding of the underlying biological concepts as they pertain to cucurbits, and a jumping-off point from which to pursue investigations on particular crop species.

Although diverse, most cucurbits do share a collection of characteristics (e.g. rapidly growing vines with tendrils, possessing relatively large fruits, adaptable to the point of becoming weeds, containing various bioactive compounds) that make them a unique, fascinating, and useful family of plants. Continued research should lead to their enhanced exploitation and appreciation.

Speaking of appreciation, we would like to thank the following individuals and institutions for their support of this book: Terrence W. Walters (fellow cucurbit researcher, husband of the second author, and reviewer of various early drafts), the American Society for Horticultural Science (for permission to use so many of their published tables and figures), and Tom Andres (for the use of material from his unpublished doctoral thesis). The first author wishes to express his appreciation to his former professors, Charley Rick and Henry Munger, for their training and inspiration for vegetable crops research. We dedicate this book to the memories of Liberty Hyde Bailey and Tom Whitaker, who did so very much to increase our knowledge and appreciation of cucurbits.

WHAT ARE CUCURBITS?

DISTRIBUTION AND ECOLOGY

The *Cucurbitaceae* is a family of frost sensitive, predominantly tendril-bearing vines which are found in subtropical and tropical regions around the globe. The few species that are native to or cultivated in temperate climates are prolific seed-producing annuals, perennials that live for one season until killed by frost, or xerophytic perennials whose succulent underground parts survive the winter. Ecologically, the family is dichotomous; many genera consist of aggressive climbers which flourish in the humid tropics, particularly in southeastern Asia and the neotropics, whereas other genera are native to the arid regions of Africa, Madagascar and North America. Members of the latter group, the xerophytes, typically have large, perennial roots and succulent stems that are clambering and creeping and at least partially subterranean; in some cases, tendrils or leaves are lacking or greatly modified.

Although most crops in the *Cucurbitaceae* have been selected from the mesophytic annuals, concern over famine and fuel sources in arid countries has led to interest in turning some of the xerophytes into agricultural crops. For example, *Acanthosicyos*, which has already served as a wild food (fruits, seeds) in Africa for 8000 years, is being studied as a firewood substitute (roots). The large-rooted buffalo gourd (*Cucurbita foetidissima* HBK) has been proposed as a source of vegetable oil and livestock feed.

NOMENCLATURE

'Cucurbits' is a term coined by Liberty Hyde Bailey for cultivated species of the family *Cucurbitaceae*. During this century, the term has been used not only for cultivated forms, but also for any species of the *Cucurbitaceae*, and it will be so used in this book.

Other vernaculars applied to the family and various of its members are 'gourd', 'melon', 'cucumber', 'squash', and 'pumpkin'. Of these, squash and pumpkin are the most straightforward, almost always referring to species of *Cucurbita*. An exception is the fluted pumpkin, which is *Telfairia occidentalis*. The unqualified terms melon and cucumber usually define *Cucumis melo* and *C. sativus*, respectively. However, confusion develops when modifiers are added to these terms. Whereas muskmelon refers to a specific type of *C. melo*, watermelon is *Citrullus lanatus*, and bitter melon indicates *Momordica charantia*. Gourd generally is used to describe a cucurbit fruit with a hard, durable rind; usually it refers to bottle gourd (*Lagenaria siceraria*), a wild species of *Cucurbita* or an ornamental form of *C. pepo*. However, various other cucurbits also are called gourds, including some that do not have hard rinds (e.g. *Coccinia grandis*, which is the ivy gourd). Sometimes the term refers to tough-rinded species of other plant families, such as the tree gourd (*Crescentia cujete* L., Bignoniaceae).

Complicating matters further, more than one common name is often applied to a single species. For example, names for *Benincasa hispida* include ash gourd, white pumpkin, wax gourd and winter melon. Sometimes different names refer to distinct crops within a species, such as pumpkin, zucchini and acorn squash in *Cucurbita pepo*. In other cases, the different common names for the same species are used interchangeably. Frequently used common names for cucurbit crops are given in the Appendix.

The vernacular conundrum is not a problem restricted to the English language. The Chinese 'guo-kua', which is translated into English as 'melon', includes watermelon (*Citrullus lanatus*) as well as melon (*Cucumis melo*); 'tsai-kua', which usually translates as 'gourd', refers to *Benincasa hispida*, *Cucumis sativus*, *Momordica charantia* and species of *Cucurbita* and *Luffa*. Unfortunately, China's agricultural statistics are often based on these two major groupings instead of on individual genera or species.

TAXONOMY

The *Cucurbitaceae*, which is not closely related to any other plant family, consists of two well-defined subfamilies, eight tribes representing varying degrees of circumscriptive cohesiveness, and about 118 genera and 825 species (Jeffrey, 1990). The four major cucurbit crops (watermelon, cucumber, melon, squash) and five other important crops (loofah, bottle gourd, chayote, wax gourd, bitter melon) in the family belong to the *Cucurbitoideae* subfamily. Four of these – watermelon, loofah, bottle gourd and wax gourd – belong to the tribe *Benincaseae*. The classification of these and other cultivated species is given in Table 1.1. Many more wild taxa have actual or potential economic value, making the *Cucurbitaceae* one of the most important plant families for human exploitation.

Systematic studies of cucurbits at all hierarchial levels are diverse. They

Table 1.1. Taxonomy of cultivated cucurbit species.

Latin name	Common name[1]	Frequency and place of cultivation[2]	Usage
Subfamily *Zanonioideae*			
Tribe *Zanonieae*			
Subtribe *Fevilleinae*			
Fevillea cordifolia L.	Antidote vine	Rare (N)	Medicinal
Subtribe *Gomphogyninae*			
Gynostemma pentaphyllum (Thunb.) Mak.	Jiao-gu-lan	Localized (S)	Medicinal
Hemsleya amabilis Diels	Luo-guo-di	Localized (S)	Medicinal
Subtribe *Actinostemmatinae*			
Actinostemma tenerum Griff.	He-zi-cao	Localized (S)	Medicinal
Bolbostemma paniculatum (Maxim.) Franq.	Pseudo-fritillary	Localized (S)	Medicinal
Subfamily *Cucurbitoideae*			
Tribe *Melothrieae*			
Subtribe *Cucumerinae*			
Cucumeropsis mannii Naud.	White-seeded melon	Localized (A)	Food
Cucumis anguria L.	Bur gherkin	Localized (W)	Food
Cucumis dipsaceus Ehrenb. ex Spach	Teasel gourd	Sporadic (W)	Ornamental
Cucumis melo L.	Melon	Common (W)	Food
Cucumis metuliferus E. Mey. ex Naud.	African horned cucumber	Sporadic (W)	Food
Cucumis sativus L.	Cucumber	Common (W)	Food
Tribe *Joliffieae*			
Subtribe *Thladianthinae*			
Momordica angustisepala Harms	Sponge plant	Rare (A)	Utilitarian
Momordica balsamina L.	Balsam apple	Frequent (W)	Medicinal
Momordica charantia L.	Bitter melon	Common (W)	Food, medicinal
Momordica cochinchinensis (Lour.) Spreng.	Cochinchin gourd	Sporadic (W)	Medicinal
Momordica cymbalaria Frenzl. ex Hook. f.	–	Rare (O)	Food

Table 1.1. continued

Latin name	Common name[1]	Frequency and place of cultivation[2]	Usage
Momordica dioica Roxb. ex Willd.	Kaksa	Localized (S)	Food
Siraitia grosvenorii (Swingle) Lu & Zhang	Luo-han-guo	Localized (S)	Medicinal
Thladiantha dubia Bunge	Red hail stone	Sporadic (W)	Medicinal, ornamental
Subtribe *Telfairiinae*			
Telfairia occidentalis Hook. f.	Fluted pumpkin	Localized (A)	Food
Telfairia pedata (Sims) Hook.	Oyster nut	Localized (A)	Food
Tribe *Trichosantheae*			
Subtribe *Hodgsoniinae*			
Hodgsonia macrocarpa (Bl.) Cogn.	Lard plant	Infrequent (S)	Food
Subtribe *Trichosanthinae*			
Gymnopetalum cochinchinense (Lour.) Kurz	–	Rare (S)	Food
Trichosanthes cucumerina L.	Snake gourd	Frequent (W)	Food
Trichosanthes dioica Roxb.	Pointed gourd	Localized (S)	Food
Trichosanthes kirilowii Maxim.	Chinese snake gourd	Localized (S)	Medicinal
Trichosanthes lepiniana (Naud.) Cogn.	Indreni	Localized (S)	Medicinal
Trichosanthes ovigera Blume	Japanese snake gourd	Frequent (S)	Food
Trichosanthes villosa Bl.	Mi-mao-gua-lou	Localized (S)	Food
Tribe *Benincaseae*			
Subtribe *Benincasinae*			
Acanthosicyos horridus Welw. ex Hook. f.	Inara	Localized (A)	Food
Benincasa hispida (Thunb.) Cogn.	Wax gourd	Frequent (W)	Food
Bryonia alba L.	Bryony	Sporadic (W)	Medicinal
Bryonia cretica L.	Bryony	Localized (O)	Medicinal
Bryonia dioica Jacq.	Bryony	Sporadic (W)	Medicinal
Citrullus colocynthis (L.) Schrad.	Colocynth	Sporadic (W)	Medicinal
Citrullus lanatus (Thunb.) Matsum. & Nakai.	Watermelon	Common (W)	Food

Species	Common name	Location	Use
Coccinia abyssinica (L.) Cogn.	–	Localized (A)	Food
Coccinia grandis (L.) Voigt	Ivy gourd	Sporadic (W)	Food
Diplocyclos palmatus (L.) C. Jeffrey	Lollipop climber	Localized (O)	Ornamental
Ecballium elaterium (L.) A. Rich.	Squirting cucumber	Sporadic (O)	Ornamental
Lagenaria siceraria (Molina) Stand.	Bottle gourd	Common (W)	Utilitarian, ornamental
Praecitrullus fistulosus (Stocks) Pang.	Round melon	Localized (S)	Food
Subtribe *Luffinae*			
Luffa acutangula (L.) Roxb.	Angled loofah	Frequent (W)	Food
Luffa cylindrica (L.) M. J. Roem.	Smooth loofah	Common (W)	Utilitarian, food
Tribe *Cucurbiteae*			
Subtribe *Cucurbitinae*			
Cayaponia kathematophora R. E. Schult.	–	Rare (N)	Ornamental
Cayaponia ophthalmica R. E. Schult.	–	Rare (N)	Medicinal
Cucurbita argyrosperma Huber	Squash, pumpkin	Localized (W)	Food
Cucurbita ficifolia Bouché	Malabar gourd	Localized (W)	Food
Cucurbita maxima Duch. ex Lam.	Squash, pumpkin	Common (W)	Food
Cucurbita moschata Duch. ex Poir.	Squash, pumpkin	Common (W)	Food
Cucurbita pepo L.	Squash, pumpkin	Common (W)	Food
Sicana odorifera (Vell.) Naud.	Casabanana	Sporadic (N)	Food
Tribe *Sicyeae*			
Subtribe *Cyclantherinae*			
Cyclanthera brachybotrys (Poepp. & Endl.) Cogn.	–	Localized (N)	Food
Cyclanthera explodens Naud.	–	Localized (N)	Food
Cyclanthera pedata (L.) Schrad.	Stuffing cucumber	Sporadic (W)	Food
Echinocystis lobata (Michx.) Torr. & Gray	Wild cucumber	Sporadic (W)	Ornamental
Subtribe *Sicyinae*			
Sechium edule (Jacq.) Swartz	Chayote	Common (W)	Food
Sechium tacaco (Pitt.) C. Jeffrey	–	Localized (N)	Food

[1]For most species, the common name given is English or an English translation. Common names in other languages are listed in Kays and Silva Dias (1995).

[2]Locations: A=Africa, N=neotropics, O=Old World, S=Asia, W=widespread.

include comparative analyses of morphology (including specialized studies on trichomes, stomata, palynology, and seed coat anatomy), cytology, DNA, isozymes, flavonoids, cucurbitacins, amino acids and fatty acids in seeds, biogeography and coevolving insects. There is a recent monograph for *Cucumis* (Kirkbride, 1993), and similar treatments are needed for other crop genera to aid crop improvement and germplasm conservation.

MORPHOLOGY AND ANATOMY

Seedlings

Most cucurbit seedlings are epigean, germinating with the tips of the cotyledons initially inverted but later erect. In cucurbits such as squash, the hypocotyl straightens and the cotyledons ascend as the seed coat is dislodged by the peg, an outgrowth on one side of the hypocotyl. The function of the peg is to open the seed coat and permit the cotyledons to emerge. The photosynthetic cotyledons of most cucurbit seedlings are more or less oblong in shape. Between them lies the inconspicuous developing epicotyl.

Roots

Cucurbits generally have a strong tap root, which may penetrate the soil for a depth of 1 m in cucumber and 2 m in squash. They also have many secondary roots occurring near the soil surface. In fact, most roots are in the upper 60 cm of the soil. Lateral roots extend out as far as, or farther than, the above-ground stems. The cortex of the primary root is apparently involved in the development of secondary roots in cucurbits. Adventitious roots may arise from stem nodes in squash, loofah, bitter melon and other cucurbits, sometimes without the stem having contact with the soil or other substrate.

Some xerophytic species have massive storage roots which enable the plant to survive severe drought. Those of *Acanthosicyos* can reach up to 15 m in length. The central taproot on one buffalo gourd plant weighed 72 kg (Dittmer and Talley, 1964). The above-ground parts of this species may die from lack of water or in response to freezing temperatures, but the plant regenerates from surviving stem tissue at the root–shoot transition area when favourable conditions resume.

Older vessels in the secondary xylem are often plugged with tyloses, especially in watermelon. The sieve tubes in the secondary phloem are among the largest found in angiosperm plants.

Stems

The herbaceous or sometimes softly woody stems are typically prostrate, trailing, or climbing, angled in cross-section, centrally hollow, sap-filled and branched. Primary and secondary branches can reach 15 m in length. Bush forms of squash have much shorter internodes and stem lengths than vining cultivars.

Many of the xerophytic cucurbits are true caudiciforms, that is, the lower part of the perennial stem, which is usually subterranean or at ground level, is thickened and drought resistant. In *Marah*, the large underground tubers are of hypocotyl and stem base origin (Stocking, 1955). The succulent stems of *Ibervillea sonora* (S. Wats.) Greene can continue to sprout new growth annually during periods of drought lasting many years. The parenchymatous storage tissue of tuberous stems and roots develops via anomalous secondary growth.

The vascular bundles of cucurbit stems are bicollateral (phloem to the inside and outside of the xylem), discrete, usually ten, and arranged in two rings around the pith cavity. The relatively large sieve tubes are also scattered in the cortex in some cucurbits (e.g. squash), serving to join all phloem elements together. The anomalous stem anatomy of cucurbits and other vines may serve to increase stem flexibility, to facilitate nutrient transport, to promote healing of injuries, or to provide protection against stem destruction via redundancy (Fisher and Ewers, 1991).

Many cucurbits have soft to rough hairs on their stems and foliage, whereas chayote, smooth loofah, stuffing cucumber and some other cucurbits are glabrous or nearly so. Trichome morphology is quite variable: hairs are glandular or eglandular, unicellular to multicellular, and simple or branched.

Leaves

Cucurbit leaves are usually simple (i.e. not divided into leaflets), palmately veined and shallowly to deeply three- to seven-lobed. There is typically one leaf per stem node. Along the stem, leaves are helically arranged with a phyllotaxy of 2/5; in other words, there are two twists of the stem, which segment contains five leaves, before one leaf is directly above another. This means that the angle of divergence between neighbouring leaves is 144° (2/5 of 360°).

Leaf stomata are mostly anomocytic, lacking subsidiary cells. The petiole in cross-section often has a crescent or ring of unequal vascular bundles, the larger ones bicollateral. Stipules at the base of the petiole are typically absent, but have been transformed into photosynthetic thorns in *Acanthosicyos*. Extrafloral nectaries, which frequently attract ants, sometimes occur on cucurbit leaves (e.g. ivy gourd).

Succulent leaves are rare, even among the xerophytic cucurbits. Those of

Xerosicyos have large water-storage cells in the inner mesophyll and perform crassulacean acid metabolism (CAM). However, the deciduous leaves of *Seyrigia* only perform C_3 photosynthesis even though the succulent stem performs CAM metabolism. The large, ephemeral leaves of most cucurbits adapted to an arid environment avoid heat damage by maintaining high levels of transpirational cooling (Rundel and Franklin, 1991).

Tendrils

Most cucurbits have solitary tendrils at their leaf axils. Tendrils are unbranched in species such as cucumber and branched in loofah (Fig. 1.1) and other taxa. They are often coiled, helping plants to cling to trellises and other

Fig. 1.1. The coiling, five-parted tendrils of angled loofah.

supports. Terminal adhesive pads develop on the tendrils of several species, allowing attachment to tree trunks and other large textured objects. Some cucurbits lack tendrils, e.g. squirting cucumber and bush cultivars of summer squash.

Tendrils in most of the cucurbit crops are interpreted as modified shoots. However, in loofah and other species, they are considered a stipule-stem complex. There are still other interpretations concerning the anatomical origins of cucurbit tendrils, and research is ongoing.

Flowers

Many cucurbits have large, showy flowers which attract pollinating insects, but *Echinocystis*, *Sechium* and some other genera have small, rather inconspicuous flowers. The typically unisexual flowers occur in leaf axils, either alone or in inflorescences. They are often white or yellow, but may be red (e.g. *Gurania*) or other colours. The hypanthium, of receptacle or fused calyx and corolla origin, is cup-shaped to bell-shaped. The sepals or sepal lobes, typically numbering five, and the corolla, which is usually five-lobed and more or less fused, extend beyond the hypanthium. Flowers have radially symmetrical, campanulate corollas which may differ between male and female flowers.

Staminate and pistillate flowers on monoecious cucumber and squash plants are originally bisexual, with both stamen and pistil primordia initiated. During ontogeny, depending on the hormonal status of the tissue near the floral bud, development of the anthers may be arrested and a pistillate flower develops, or development of the pistil is retarded and a staminate flower is produced. Undeveloped stamens (staminodia) can be seen in mature pistillate flowers, and there is a rudimentary pistil (pistillodium) in staminate flowers (Fig. 1.2).

Stamens are attached to the hypanthium and alternate with corolla lobes. The basic number of stamens in the *Cucurbitaceae* is five. Some cucurbits (e.g. *Fevillea*) have five free stamens, whereas all five stamens are fused together in *Cyclanthera*. During evolution, fusion of two pairs of stamens has resulted in some genera (e.g. *Cucumis*) having one small, unilocular and two large, bilocular stamens. The three stamens are usually attached to each other to some degree by their anthers, as in squash; filaments are short and often united also.

Pollen grain morphology is relatively homogeneous in the *Zanonioideae* and quite variable in the *Cucurbitoideae*, but consistent within genera. In squash, pollen grains are very large, spherical and spiny.

Pistillate flowers have an inferior ovary below the hypanthium. The pistil, which often has a fused style but separate stigma lobes, generally has three or five carpels. The fleshy placentae bear numerous ovules in most species, but only one in chayote.

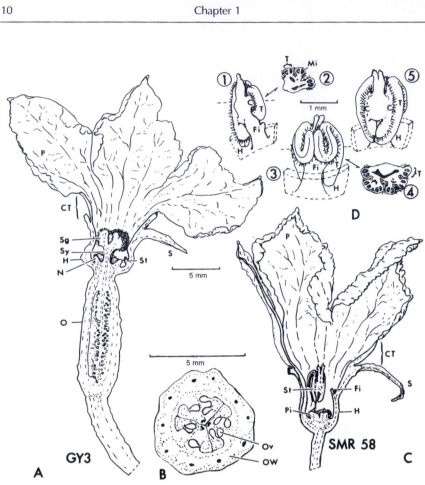

Fig. 1.2. Comparison of female and male cucumber flowers. (A) Female flower at anthesis, longitudinal section. (B) Tricarpellate ovary, transection at anthesis. (C) Male flower at anthesis, longitudinal section. (D) Simple stamen (1, 2) and two compound stamens (3–5). C, anther connective; CT, corolla tube; Fi, filament; H, hypanthium; Mi, microsporangium; N, nectary; O, ovary; Ov, ovule; OW, ovary wall; P, petal; Pi, pistillodium; S, sepal; Sg, stigma; St, stamen or staminodium; Sy, style; T, theca. (Goffinet, 1990. Reprinted courtesy of Cornell University Press.)

Floral nectaries attract pollinating insects. These structures are borne inside and at the base of the hypanthium in both staminate and pistillate flowers. The nectary forms a continuous ring surrounding the base of the style(s) in the female flower (Fig. 1.3), whereas the nectary and its associated pistil rudiment form a button-shaped mound at the centre of the male flower.

Fruits

Fruits of the *Cucurbitaceae* are extremely diverse in many characteristics, including size, shape, colour and ornamentation. Those of bryony are small (ca 5 mm), spherical, and green, red or black in rind colour. Angled loofahs are club-shaped, about 60 cm long, and prominently ribbed. Some of the many shapes of bottle gourds are described in Chapter 4. The striped, mottled, bicoloured or solid-coloured fruits of squash are smooth, wrinkled, warted, furrowed or ridged. The stiff, dry spines of *Sicyos* facilitate animal dispersal, whereas those on the teasel gourd are soft and fleshy.

Some cucurbit fruits, especially those of squash, are the giants of the plant kingdom. *Cucurbita maxima* is well named, for it is in this species that maximum fruit size in the genus occurs. Every year, there is a contest to grow the world's largest pumpkin. In this contest any squash fruit with an orange skin is considered a pumpkin, and the winners are invariably *C. maxima*. In 1995, the winning fruit weighed an astounding 440 kg.

Cucurbit fruits are generally indehiscent 'pepos', usually with one or three ovary sections or locules (Fig. 1.4). A pepo is a fleshy fruit with a leathery, non-septate rind derived from an inferior ovary. However, fruits of

Fig. 1.3. Bee feeding on nectar from the nectary ring at the base of the styles in a female squash (*C. moschata*) flower. Although squash flowers typically have united styles, each of the three bilobed stigmas is subtended by a completely separate style in this cultivar from China.

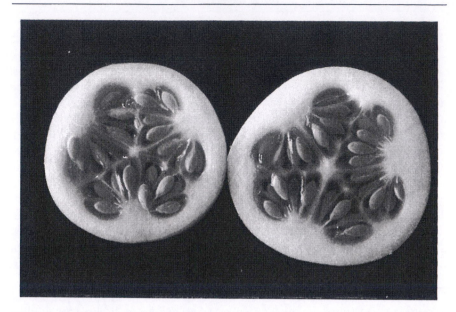

Fig. 1.4. Cross-section of two fruits of a Dudaim Group melon, each measuring about 3 cm in diameter. Note parietal placentation of seeds in the tricarpellate pepo.

some cucurbit genera, e.g. *Momordica* and *Cyclanthera*, split at maturity. Fruits of squirting cucumber forcefully eject their seeds through a blossom-end pore. Fruits may be dry when mature, as in loofah, where seeds fall out through a hole at the bottom of the pendulous fruit. Mature fruits of many cucurbits have a hard, lignified rind, but various squash cultivars have been bred to have a tender rind.

Fruit flesh is generally white to pale yellow and moist. Cultivars of melon and squash have been bred to have orange flesh, due to carotenoids, and most watermelon cultivars contain lycopene, which produces red flesh. In melon, watermelon and squash, the flesh is derived from the fruit wall. In other species, including cucumber, the edible flesh may be mostly placental in origin.

When a loofah fruit matures and dries, what is left is a papery outer skin and a fibrous mass surrounding the seeds. This tangled mass of modified vascular bundles, consisting mostly of fibre cells, makes up a loofah sponge.

Seeds

There may be only one seed in the fruit, as in chayote, or more typically, tens to hundreds of seeds. Cucurbit seeds, which are rarely winged, are usually flat.

The seed coat encloses a collapsed perisperm, an oily embryo and little or no endosperm. The endosperm is consumed during seed development. Two cotyledons make up much of the contents of the seed. Seeds of some cucurbits are enveloped in a false aril of placental origin; in bitter melon this sarcotesta is red and fleshy, attracting birds as seed dispersal agents.

Seed size, shape and colour vary greatly among the cultivated cucurbits (Fig. 1.5). The largest unwinged seed is that of *Hodgsonia*, measuring about 7 cm long. The nearly spherical seed of *Bryonia* is sometimes less than 3 mm in diameter. Cultivars within a crop, including watermelon and squash, may differ considerably in their seed sizes and other seed characteristics. Depending on the cultivar, watermelon seeds are black, black with brown or tan centres, brown with black stippling, golden brown, red, olive green or tan to white, with or without a few black, marginal markings.

The complex seed coat anatomy of cucurbits has been well studied (Singh and Dathan, 1990). The testa develops from the outer integument of the anatropous, bitegmic, crassinucellate ovule. The inner integument degenerates in fertilized ovules. The mature seed coat in the subfamily *Cucurbitoideae* consists of an epidermis, hypodermis, main sclerenchymatous zone, aerenchymatous zone, and inner parenchymatous or chlorenchymatous zone. In the *Zanonioideae*, the sclerenchymatous layer is poorly or not differentiated from

Fig. 1.5. Seeds of the *Cucurbitaceae*. The largest seed (*Fevillea cordifolia*) shown here measures ca 53 mm across. Also shown are seeds of *Cucurbita, Siraitia, Marah, Momordica, Sicana, Trichosanthes, Luffa, Lagenaria, Echinocystis* and *Melothria* (the smallest seeds).

the hypodermis, and the well-developed aerenchyma has distinctive lignified thickenings. Within each subfamily, there is further anatomical diversity among genera.

GROWTH AND DEVELOPMENT

Seed Germination

Seed maturation usually continues until the fruit starts to yellow with senescence. Some seed producers store mature cucurbit fruits after harvest to permit the seeds to develop further. However, if seeds are left too long in some fruits, they may germinate *in situ*. In a few cucurbits, such as chayote, germination is naturally viviparous (Fig. 4.17).

Seed dormancy, which is common in various wild species, is not usually a serious problem in the major crops. Dormancy can occur in freshly harvested seeds of some cultivars, but this dormancy can be broken by a month or more of after-ripening, i.e. storing seeds in the fruit after harvest. Light and low temperature (<15°C) are strong inhibitors of germination for many species. Under amenable conditions (e.g. low light levels, temperatures of 25–30°C and adequate but not soaking moisture), germination takes 2 days to 2 weeks if the seeds are not dormant. See Chapter 5 for experimental studies investigating seed germination physiology.

Plant Growth and Movement

Most cucurbits grow very rapidly in warm weather, with stem extension growth outpacing leaf development in the tuberous perennials. In a single growing season, a wild buffalo gourd plant produced 360 shoots covering an area 12 m in diameter with a total vine length of over 2000 m (Dittmer and Talley, 1964). Among the annuals, bottle gourd stems can elongate up to 60 cm in 24 h, and wild cucumber (*Echinocystis lobata*), which is adapted to the short growing seasons of southern Canada, is considered one of the fastest growing vines. Holroyd (1914) reported that a single annual squash (*Cucurbita pepo*) plant produced 450 leaves on a vine measuring 43 m. Cucurbit root growth is also rapid, occurring at a rate of 6 cm per day for squash when conditions are favourable.

In large-fruited, vining squash plants, total leaf area typically increases exponentially throughout the season until fruit set creates a large repro-ductive sink and vegetative growth is suppressed. During the period between flower primordia initiation and the start of fruit development in bur gherkin, differentiation of new vegetative organs decreases, the growth rate of existing vegetative organs increases, and water and nutrient intake drops; soon after fruit development begins, vegetative differentiation and water and nutrient intake resurge (Hall, 1949).

Biomass productivity differs among species and cultivars, and is influenced by cultural practices (e.g. planting density, irrigation, fertilizer application) as well as local environmental conditions. Environmental factors most affecting growth rates are photoperiod and ambient temperature, either of which can affect the intake and effective utilization of water and nutrients. Many investigations of the relationship between growth and environmental factors have been carried out on species of *Cucumis*, particularly cucumber. The results of these studies, which are discussed in detail in Whitaker and Davis (1962) and Wien (1997), are summarized below.

1. The growth rate curve for a single leaf under continuous light is generally an S curve, but is affected by light intensity.
2. The rate of stem elongation is greater during 8-h days than during 16-h days, and plants grown under short-day conditions produce more nodes and leaves, but smaller total leaf and root areas.
3. Overall stem length may be greater under a long-day versus a short-day regime when nitrogen levels are high.
4. Under low-nitrogen conditions, plants grown during long days contain more carbohydrates at anthesis than plants grown during short days, but this carbohydrate relationship is reversed at fruit maturity.
5. Stem extension and leaf area growth rates are linearly dependent on mean ambient temperature during periods of optimum temperatures for growth (20–30°C, depending on other environmental conditions). However, in the range of 15–27°C, Grimstad and Frimanslund (1993) found that plant dry weight of cucumber had a sigmoidal response curve with inflections at about 17 and 24°C.
6. When temperature rises above the optimum, leaf growth rate in young plants declines as material is redistributed to the stems and cell division in developing leaves is reduced.
7. At below-optimum temperatures, relative leaf growth rate is independent of temperature and controlled instead by light intensity.
8. Stem extension rates are lower than normal when night temperatures exceed day temperatures.
9. Low temperatures slow down development of apical buds.

In *Cucurbita pepo*, bushy plants with short internodes possess an allele that reduces biosynthesis of endogenous gibberellin. When these plants are treated with high concentrations (2.9–4.3 mmol l^{-1}) of gibberellic acid, internode lengths become as long as those of naturally viny squash plants (Fig. 1.6).

Breeders have also selected for bush cultivars in *C. maxima*. Research on bushy versus viny plants of this species indicates that bush cultivars have a more uniform growth pattern, respond better to high-density planting, and produce a greater percentage of fruit versus vegetative biomass during short growing seasons (Loy and Broderick, 1990). This last effect is partly due to the fact that fruiting begins sooner in bush cultivars, which in turn, suppresses vegetative growth. However, photosynthetic rates in bush plants increase during fruit set in response to increasing sink demand, which may be possible because of a proportionally thicker palisade layer in bush plant leaves (Loy and Broderick, 1990).

In climbing cucurbits, the stems often revolve, twist, and extend upwards. Darwin (1906) made several interesting observations on the revolving nature of cucurbit stems and tendrils. He noted that the average rotation rate was 100 min per revolution in wild cucumber. Light affects this movement, with stem tips, including the two uppermost internodes subtending the apical meristem, following the sun throughout the day.

Generally, the slightly curved tendril becomes sensitive to mechanical stimulus on its concave side when it is one-third grown. At that stage, it reacts quickly (in under 2 min), with coiling caused by the elongation of parenchymatous cells on the convex side of the tendril. The revolving movement of a tendril does not stop after it has coiled, but its ability to coil is limited after it stops revolving.

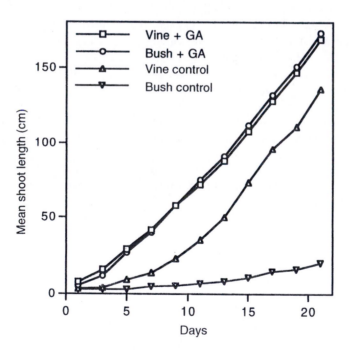

Fig. 1.6. The effect of 2.9 mmol l⁻¹ gibberellic acid (GA) on shoot elongation of bushy and viny lines derived from 'Table Queen' (*Cucurbita pepo*) grown under glasshouse conditions in the winter. (Redrawn from Denna, 1963, with permission from the American Society for Horticultural Science, 600 Cameron Street, Alexandria, VA, 22314, USA.)

Sex Expression

No cucurbit species is known to have only or primarily functionally hermaphrodite flowers. Instead, most cucurbits are monoecious, that is, they have separate male and female flowers on the same plant. Among its genera, the *Cucurbitaceae* also has a high rate of dioecy, where staminate and pistillate flowers occur on separate plants. Some genera have only dioecious species. Half of the 135 cucurbit species surveyed in India by Roy and Saran (1990) were dioecious, a much higher proportion than is typical for angiosperm families. Cultivated cucurbits that are dioecious include oyster nut, fluted pumpkin, ivy gourd and pointed gourd.

Primitive cucurbits are typically monoecious, as are the majority of domesticated cucurbits. Most squash and watermelon cultivars are monoecious, although genes for different forms of sex expression are known for these crops. For example, some watermelon cultivars may have three types of flowers on the same plant: staminate, pistillate and hermaphrodite. Many cucumber cultivars are monoecious, but others are gynoecious or predominantly female and 'Lemon' cucumber is andromonoecious, meaning that it has male and bisexual flowers on the same plant. Angled loofah is monoecious except for the cultivar 'Satputia', which has only hermaphrodite flowers. Most melon cultivars are andromonoecious. The bisexual flowers of these plants are often borne on the first or second node of the lateral branches. 'Banana' and other melon cultivars of the Flexuosus Group are monoecious.

Monoecy is the ancestral condition in cucurbits. Dioecy and other forms of sex expression have arisen in various evolutionary lines in the family. Single genes can determine the occurrence of unisexual plants in normally monoecious species, as in the case of androecy (all male flowers on a plant) in *Cucurbita pepo*. In melon, cucumber and many other cucurbits, two or more genes are involved in sex expression, sometimes with each gene having three or more alleles (e.g. loofah). The development of heteromorphic sex chromosomes (e.g. ivy gourd, see Chapter 3 for more details) to determine dioecy is considered to represent the ultimate degree of evolution from monoecy in the *Cucurbitaceae*.

Monoecious cultivars may differ in degree of female sex expression, some having a higher proportion of female to male flowers. Generally, they produce many more staminate than pistillate flowers and go through a progression of floral development. Nitsch *et al.* (1952) determined that young squash plants are initially vegetative, then bear only underdeveloped male flowers. Later, they produce only normal male flowers, then bear normal female as well as male flowers. The proportion of female to male flowers increases as the plants grow older, and the plant eventually produces only female flowers. If squash plants are not pollinated, ultimately they may produce enlarged female flowers and parthenocarpic fruits.

Although staminate flowers usually appear a few days to a few weeks before pistillate flowers on squash vines, a few cultivars and some wild populations of *Cucurbita pepo* may produce female flowers first (Fig. 1.7), especially when spring temperatures are low. These wild populations also tend to have a higher ratio of female to male flowers than most cultivars (Decker, 1986).

In their search for early staminate flower production, Walters and Wehner (1994) examined 866 cucumber cultivars, breeding lines and other accessions. Earliness was normally distributed for the cultigens, and ranged from 26 to over 45 days from planting to first staminate flower.

Sex expression in the *Cucurbitaceae* is controlled by environmental as well as genetic factors. Unfavourable growing conditions (e.g. lack of water) can cause a slow down in flower production altogether. In general, female sex expression is promoted by low temperature, low nitrogen supply, short photoperiod and high moisture availability, i.e. conditions which encourage the buildup of carbohydrates. These environmental factors influence the levels of endogenous hormones (especially ethylene, auxin and gibberellic acid), which in turn, influence sex expression. Most studies examining this role of hormones have been conducted on cucumber, followed by squash, melon and

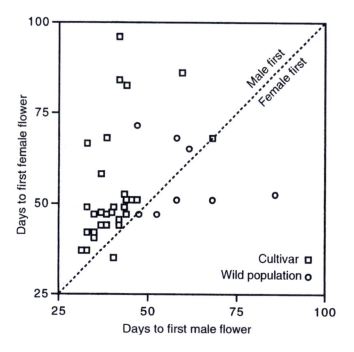

Fig. 1.7. Accession means of the number of days from seed germination to the first male and female flowers on plants of 30 cultivars and eight wild populations of *Cucurbita pepo*. (Data from Decker, 1986.)

watermelon. See Chapter 6 for details on the use of exogenous hormones to control sex expression.

Temperature affects anthesis and, in squash, the length of time that a flower is open. Squash pollen is released at temperatures as low as 10°C, whereas cucumber, watermelon and melon flowers require higher temperatures for anther dehiscence. Warmer temperatures cause anthesis in squash to occur earlier in the morning. However, high temperatures (ca 30°C) accelerate squash flower closing, causing the corollas to close by mid- to late-morning.

Fruit Development

Pollen tube growth and ovule fertilization stimulate ovary enlargement. Subsequent fruit set depends on the quality of pollination (i.e. having enough ovules fertilized) and is affected by the presence of already developing fruits, leaf area, daylength and other environmental factors. Fruits on a plant may inhibit the production of additional pistillate flowers and the development of subsequent fruits. In cucurbit crops such as melon, watermelon and squash, fruit-thinning will allow fruits that remain on the vine to grow larger.

If fruits are not developing on a plant by the end of the growing season, then the last group of ovaries may develop parthenocarpically. Parthenocarpy in cucumber and squash is promoted by low temperature, short daylength and old plant age. Some cucumber cultivars (e.g. those that are gynoecious) are more genetically predisposed to parthenocarpy than others.

A century ago, a curious researcher (Anderson, 1894) weighed a developing squash fruit hourly and determined that the fruit gained weight at an average rate of one gram per minute. The greatest weight increase was at night. The growth rate for a particular cucurbit fruit is influenced by exogenous conditions (e.g. higher temperatures and greater light exposure increase the growth rate) as well as by endogenous plant conditions, such as the presence of other developing fruits, which retards growth.

Several classic studies on the inheritance and development of fruit shape in squash (Cucurbita pepo) were conducted by Sinnott (1932). He found that fruit shape is more or less evident in the shape of the immature ovary, with the developmental changes that affect ultimate shape due to both genetic and environmental factors. Also, a few fruits on a plant, particularly the earliest ones, may be shaped a little differently than the rest (Fig. 1.8a), with this difference being evident in the ovaries (Fig. 1.8b).

Immature ovaries are typically green, although those of squash cultivars with gene B may be yellow. Fruits of various cucurbits, including loofah and bottle gourd, remain green until fruit senescence, at which time they turn tan or brown. Other cucurbits develop rind coloration changes during maturation. In these, chlorophyll depletion reveals the presence of additional

Fig. 1.8. (*and opposite*) Fruit development in squash (*Cucurbita pepo*). (A) From left to right, these fruits of 'Mandan' developed consecutively on the same plant. (B) The fruit shape differences in 'A' were evident in the unfertilized ovaries, similar to those shown here. (C) The pale markings on the ovary of this female flower of var. *texana* suggest the mottled green and white striped pattern that will form later as the fruit develops. (D) In 'Mandan', green stripes do not appear on the fruit until full size is nearly reached, as evidenced by this young (left) and older fruit from the same plant.

pigments after pollination. For example, the green fruits of many squash cultivars become yellow or orange as they age. Colour changes usually begin at the blossom end of the fruit. In most wild and cultivated *Cucurbita*, rind designs, such as stripes or mottling, are lightly visible on the ovary (Fig. 1.8c), becoming more distinct soon after pollination. However, the white fruits of 'Mandan' (*C. pepo*) reach almost full size before the dark green markings appear (Fig. 1.8d). Fruit markings persist at maturity for some squash cultivars, but fade away in senescent fruits of other cultivars.

In young melon fruits of the Cantalupensis Group of cultivars, rapidly dividing cork cells develop below the epicarp. Towards maturity, this growth breaks through to form a network of grey, corky tissue covering the rind, as is evident in the netting of muskmelon cultivars.

A unique case of adaptation to a particular ecological niche is exemplified by *Cucumis humifructus* Stent. This African species has a geocarpic fruit, similar to that of the peanut plant. After flowering and setting fruit above the ground, much the same as other *Cucumis* species, the developing fruit is thrust downward and completes its development several inches below the soil surface. In its native land, the African anteater or aardvark (*Orycteropus afer*) consumes the subterranean fruit. It is a symbiotic relationship, with the aardvark using the fruit as a source of food and water in the arid area where the plant often grows, and providing a means of seed dispersal for the plant. *Cucumis humifructus* is called aardvark cucumber because of this relationship, but it is much more closely related to melon than to cucumber.

2

EVOLUTION AND EXPLOITATION

EVOLUTIONARY HISTORY

'Cucurbits are weeds waiting to become crops.' While this anthropomorphic statement may be a generalized over-simplification of crop evolution in the *Cucurbitaceae*, supporting evidence can be seen in the vast array of valuable plant products found within the family. Aboriginal plant gatherers were probably attracted to some of these products, particularly the relatively large, long-keeping and sometimes showy fruits. After fruits were taken back to camp, seeds that were purposely discarded, accidentally dropped or partially digested found new life on rubbish heaps, settlement edges or other disturbed areas within the camp. Eventual recognition of the value of the resident cucurbits led to their tolerance, horticultural care and further exploitation. Finally, seeds and, more rarely, vegetative propagules were carried by and exchanged among migrating bands of these incipient cultivators, gradually turning the earliest cultivated cucurbits into domesticated crops.

It was not only because cucurbits were wanted by human gatherers that they became domesticated. Certain physiological and genetic characteristics generally associated with weeds, or colonizing species, allowed cucurbits to adapt to human habitats. Fast, indeterminate growth, developmental plasticity in response to environmental conditions (especially regarding sex expression), and genetic diversity at the genomic, chromosomal and gene levels enabled these cucurbits to continue their survival through the coevolutionary relationship called domestication. In turn, this relationship brought many changes to domesticated cucurbits: the size of various plant parts increased, bitterness was eliminated from and overall flavour and appearance improved in fruits, early female flower production was selected for and dormancy in seeds was reduced.

Given the nature of cucurbits, it is not surprising that they are among the most ancient of cultivated plants. Archaeological evidence suggests a pan-tropical distribution for bottle gourd going back more than 10,000 years.

23

Bottle gourd may have been cultivated in Asia, Africa and the New World at that time, or later if the oldest remains are of wild plants. Seeds and/or rinds of *Cucurbita pepo* have been found at sites dating 10,000–30,000 BC in Florida (Lee Newsom, Illinois, 1995, personal communication), 7000–9000 BC in Mexico and ca 5000 BC in Illinois. Enlarged fruit parts indicate that this and other squash species were being cultivated for food in the New World by 5000 BC, if not earlier. Cucumber has been cultivated in India for at least 3000 years, and both it and melon were grown by early Egyptians, Greeks and Romans. Melon was one of the most important vegetables in ancient China, with archaeological remains there dating back to 3000 BC . Given that this species is probably African in origin, original domestication may have taken place thousands of years earlier. Another African native, watermelon, was an important crop in northern Africa and southwestern Asia by 4000 BC. It is recorded in the Bible's Book of Numbers that the Jews missed the watermelons and cucumbers of Egypt during the Exodus in the twelfth century BC. A watermelon relative, colocynth, is probably the toxic fruit mentioned in 2 Kings of the Bible.

USES

Uses of cucurbits are extremely diverse. Fruits may be eaten when immature (e.g. summer squash) or mature (watermelon). They are baked (squash), fried (bitter melon), boiled (wax gourd, snake gourd), stuffed (stuffing cucumber), pickled (cucumber, pickling melon), candied (watermelon, wax gourd, Malabar gourd), or consumed fresh in salads (cucumber) or as a dessert (melon). Fresh fruits are sold soon after harvest or following a storage period of several weeks to months. Processing for long-term storage often involves canning, freezing, or pickling (Fig. 2.1). Juice from the fruits of watermelon, Malabar gourd and other cucurbits is fermented to make alcoholic beverages; e.g. a liqueur is made from melon in Japan. In addition to the fruits, seeds (*Cucurbita*, *Cucumeropsis* and many others), flowers (squash), leaves and stem tips (*Telfairia*, *Momordica* and others), and even roots (chayote) provide human sustenance.

Seeds of watermelon, oyster nut, !nara and other cucurbits are an important part of the African diet. In Mexico and Central American countries, squash seeds are sold in stores and at street corners as snacks. The development of squash and pumpkin cultivars with 'naked' seeds, devoid of tough seed coats, has increased the popularity of squash seeds for food in other countries.

Cucurbits are utilized for many other purposes besides food and drink. From bottle gourds, people have fashioned storage containers, bottles, utensils, smoking pipes, musical instruments, penis sheaths, masks, floats for fish nets, rattles for babies and other items. The dry fruit rinds of other

Fig. 2.1. Pickled snake melon from Lebanon. (Reprinted by permission from The New York Botanical Garden, © 1993, *Economic Botany* 47, 99, J. Ruh and J. Thieret, photographers, in Walters and Thieret, 1993.)

cucurbits, including wax gourd, wild and cultivated species of *Cucurbita*, and fluted pumpkin, sometimes serve as containers and utensils as well. Oil extracted from seeds of watermelon, loofah, antidote vine, lard plant, white-seeded melon and other cucurbits is prepared for cooking, illumination, candle and soap manufacture and industrial purposes. The fibrovascular system of a mature loofah fruit provides a sponge suitable for various purposes, including as a filter. Nigerians pound and work the thick stems of sponge plant to yield white, absorbent fibres which serve as a washing sponge. Stem fibres of fluted pumpkin are extracted for use as a sponge and for making rope. Jewellery is fashioned from seeds of various cucurbits (*Cayaponia kathematophora*, *Marah* species). Squash flowers are used in make-up, and cucumber fruit extracts are added to soap and shampoo. Arabs employ the dried fruit pulp of colocynth in the making of gunpowder, tinder and fuses.

Medicinal applications of cucurbits are too plentiful to enumerate. Since ancient times, indigenous cultures worldwide have employed a variety of

cucurbits to cure a plethora of ailments. A Chinese medicinal text written during the first century AD mentions the therapeutic benefits of wax gourd, pickling melon, bottle gourd and Japanese snake gourd. In modern Chinese commerce, he-zi-cao, pseudo-fritillary, jiao-gu-lan, luo-guo-di, luo-han-guo and species of *Trichosanthes* are the most important medicinal cucurbits (Yang and Walters, 1992).

Some of the reported therapeutic properties, including the use of cucurbits as purgatives, emetics, anthelmintics and vermifuges, are due to cucurbitacins, which are very bitter triterpenoid compounds known to have drastic effects on the digestive system. Cucurbitacins appear to be present in almost all cucurbits, usually throughout the plant, but especially in fruits and roots. Other toxic and potentially medicinal compounds in cucurbits include saponins, free amino acids and alkaloids.

Cucurbit fruits, and their extracts, are the plant part most commonly prepared for medicine. The gourd-like fruit of colocynth is used as a purgative for constipation and worms, as an antitumour agent against cancer, and as a remedy for fever, urogenital disorders and other problems. The bitter roots of many cucurbits (e.g. bryony) have cathartic effects as well. Leaves are employed for medicine less frequently, often dried and pounded into a powder or made into a tea (e.g. jiao-gu-lan). Seeds of squash, wax gourd, bottle gourd and other cucurbits are taken as anthelmintic medicine. Oil extracted from seeds of antidote vine is used medicinally to counteract the poison of a snakebite, to ward off dandruff and for a variety of other ailments.

One of the most widely employed of the medicinal cucurbits is bitter melon. Research has shown that this plant serves to control, though not cure, mild to moderately chronic cases of diabetes mellitus by increasing carbohydrate utilization. Medical investigators in China, Japan and India are testing species of *Momordica* for their purported analgesic, abortifacient, immunosuppressive and antitumour properties. Similar laboratory research is ongoing for species of *Trichosanthes*. In 1989, 'compound Q', which is an active principle extracted from Chinese snake gourd, received a lot of attention when medical professionals began to study it as a treatment against the human immunodeficiency virus (HIV). Bitter melon has also been used to combat HIV infection.

The bioactive compounds in cucurbits also serve other purposes. Mexicans have long taken advantage of the saponins in fruits of wild *Cucurbita* for creating a frothy cleansing soap. Similarly, Nigerians use the leaves of balsam apple to clean metals, and the leaves and fruits to wash the body. Saponins in seeds and fruits (e.g. loofah is used in Australia) make an effective component in fish poison. Fruits of casabanana are hung in the house as a room deodorizer. Various cucurbits are utilized to repel insects; the Chinese spray extracts of *Luffa* and *Momordica* on crops to control spider mites and other agricultural pests (Yang and Tang, 1988). Arabs smear a rind extract of colocynth on waterbags to repel camels.

Cucurbit fruits sometimes become objects of decoration and art. In the

USA, ornamental gourds and turban squashes provide autumn table decorations, and pumpkins (*Cucurbita pepo*) are carved and illuminated with candles to celebrate the holiday of Halloween. The ultimate degree of artistic expression has been reached with bottle gourds. Many different cultures independently developed the custom of carving, painting and otherwise decorating the gourds that were so useful to them. The art of gourd decoration is most highly developed in Peru, where even today, detailed pyroengraving techniques produce gourdcraft suitable for export (Fig. 4.10, p. 91).

Finally, some cucurbits are grown as ornamental plants. In addition to its lacy leaves and colourful fruits, the fragrant blossoms of bitter melon make this species a very appealing trellis ornamental; it has been grown in glasshouses in the UK since Victorian times. The beautiful white, fringed petals of snake gourd and its very long, serpent-like fruits, dangling beneath a garden arbour, make for a very picturesque sight. The spiny fruits of teasel gourd and African horned cucumber and the explosive fruits of squirting cucumber and *Cyclanthera explodens* have brought these cucurbits into commercial trade as well. The most recent horticultural trend affecting cucurbits is the current passion for succulents. Collectors are beginning to learn of and trade among themselves the caudiciform cucurbits of Africa (e.g. *Gerrardanthus* and *Trochomeria*), Madagascar (*Seyrigia* and *Xerosicyos*) and North America (*Marah* and *Ibervillea*).

COMMERCIAL PRODUCTION AREAS AND IMPORTANCE

According to the United Nations' Food and Agriculture Organization (FAO), watermelon is the world's most popular cucurbit (Table 2.1). The FAO estimated that 1,824,000 ha of watermelon were grown worldwide in 1994, producing 29,360,000 metric tonnes of fruit. Next in total world production were cucumber, melon, and then squash and pumpkin. All of these crops have been produced in greater quantities, on more land and with greater yields compared with 15 years earlier (Table 2.1). The greatest production growth (60%) has occurred for melon and is due mostly to a 35% increase in area under cultivation. In contrast, the 38% gain in cucumber production has resulted mainly from a 25% rise in yield. Glasshouse practices in The Netherlands have led to yields of over 500,000 kg ha^{-1} of cucumbers and over 45,000 kg ha^{-1} of melons (FAO, 1995).

China remains the world's leading producer of the major cucurbit crops. Although China is also a prominent producer of loofah, wax gourd and other cucurbits of Asian origin, production statistics for these crops are not readily available. China exports various cucurbit products, including fresh fruits, watermelon and squash seeds, and dried fruits of the medicinal cucurbit called

Table 2.1. Leading countries in the production of cucurbits.

Crop	Country	Area harvested (1000 ha)		Yield (kg ha^{-1})		Production (1000 tonnes)	
		1979–81	1994	1979–81	1994	1979–81	1994
Squash and pumpkin	China	70	94	10,513	17,210	737	1,616
	Ukraine	–	44	–	17,955	–	790
	Argentina	31	38	10,191	9,813	318	368
	Romania	131	80	3,215	4,500	421	360
	Turkey	25	19	14,267	17,947	367	341
	World Total	564	668	10,328	12,588	5,825	8,404
Cucumber	China	395	468	11,841	17,199	4,678	8,051
	USSR (former)	193	–	7,945	–	1,533	–
	Iran	57	96	12,217	15,729	693	1,510
	Turkey	32	40	15,968	27,500	503	1,100
	USA	68	70	12,361	14,025	834	984
	World Total	1,093	1,215	12,726	15,855	13,908	19,261
Melon	China	89	133	17,120	26,617	1,528	3,540
	Turkey	75	100	20,783	17,000	1,567	1,700
	Iran	42	86	11,182	13,779	475	1,185
	Spain	68	50	11,438	18,249	781	916
	USA	43	42	16,777	20,538	727	859
	World Total	594	803	14,554	17,307	8,648	13,894
Watermelon	China	230	363	18,387	18,623	4,228	6,760
	USSR (former)	428	–	8,828	–	3,702	–
	Turkey	151	135	20,952	25,926	3,157	3,500
	Iran	158	140	10,713	18,429	1,700	2,580
	USA	80	84	13,785	21,556	1,102	1,814
	World Total	1,786	1,824	13,990	16,095	24,973	29,360

Source: FAO (1995). Values for 1979–81 are averages over those years.

luo-han-guo (Fig. 2.2). In 1990, the Chinese Agricultural Department estimated that 30% of the land in China that was devoted to vegetable production was sown in cucurbits (Yang and Walters, 1992).

In 1994, the second leading producer of watermelons and melons was Turkey. Iran was the second leading producer of cucumbers, and Ukraine was second for squash (Table 2.1). As the fourth leading producer of melons, Spain supports its important export crop with active breeding and disease management research programmes as well as with germplasm preservation operations such as the Melon Germplasm Bank in Málaga.

It is very difficult to obtain reliable statistics on cucurbit production in many countries because governmental statistics are not readily available to

Fig. 2.2. Luohan beverage mix from China. Seeds and fruits of luo-han-guo are pictured on the package.

the public and data gathering and reporting procedures vary from country to country. Also, different agencies report different statistics for the same country. For the USA, FAO production statistics appear to be underestimates when compared with those of the United States Department of Agriculture (USDA).

Besides the major crops, there are some minor cucurbit crops of international commercial importance (Fig. 2.1). Already mentioned are the fruits of luo-han-guo exported from China. China and Japan both export other cucurbits of medicinal interest, such as preparations of *Gynostemma* (Fig. 2.3) and *Trichosanthes*, which are finding their way into the American herbal medicine market. Medicinal preparations of colocynth have a long history of trade from African to European pharmacies. Fruits of African horned cucumber that are commercially produced in New Zealand sometimes appear in American and European food markets sporting the trade name 'kiwano'. Because of expanding demand, Kenya and Isreal have also become important exporters of this attractive fruit. Japan and Central America are the major exporters of loofah sponges. In the early 1940s, when loofah sponges were an important source of industrial filters, Brazil exported almost 2,000,000 sponges per year to the USA alone. In 1985, Costa Rica was the leading exporter of chayote. Mexico and Costa Rica still ship large amounts of chayote to other Latin American countries, USA and Europe. Decorated bottle gourds are a growing source for foreign currency in Peru. Snake gourd and bitter

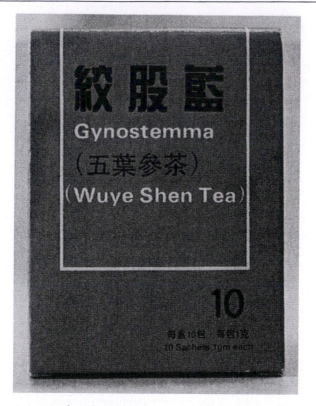

Fig. 2.3. Gynostemma tea bags from China.

melon (Fig. 2.4) are bottled or canned for export in Asia.

Cucurbits are important in the international seed trade. In addition to the major food crops, seed companies offer germplasm of other edible and ornamental cucurbits such as loofah, wax gourd, bottle gourd, bitter melon, stuffing cucumber and casabanana.

CUCURBITS AS WEEDS

Many domesticated species can become troublesome weeds. Feral populations of cucurbit crops occur along roadways, railroad tracks, streambanks, settlements and waste areas in the Americas, Europe, Africa, Asia, Australia and even islands in the Pacific. Particularly widespread and persistent are weedy forms of bitter melon, watermelon (e.g. citron) and species of *Cucumis*. Also common are escapes of species of *Cucurbita* and *Luffa*, ivy gourd, colocynth, bottle gourd, squirting cucumber and red hail stone.

Cotton fields in southern Asia and southeastern USA are plagued with

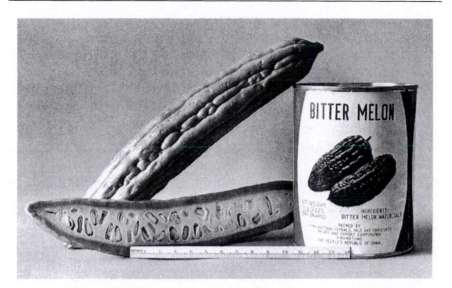

Fig. 2.4. Bitter melon, immature fruit and canned. (Reprinted by permission from The New York Botanical Garden, ©1988, *Economic Botany* 42, 287, J. Ruh and J. Thieret, photographers, in Walters and Decker-Walters, 1988.)

weedy plants of Dudaim Group melon and gourd (*Cucurbita pepo* ssp. *ovifera*), respectively. In recent years, the gourd problem has become so severe in Mississippi that cotton yields have been reduced and machinery broken.

Bur gherkin is reportedly a pest in sugar cane and peanut fields in Australia. This species also grows wild in many areas of the New World, including Brazil and various Caribbean islands. It is not native to the New World; rather, it escaped from cultivation, apparently after being introduced by African slaves.

Commercial seed trade has been another disperser of weeds. During the nineteenth century, species of Old World bryony were sold in the USA as ornamentals and sometimes escaped. Similarly, wild cucumber, a native of North America, has escaped from European gardens.

CHEMICAL COMPOSITION

Pharmacology

Toxic and potentially therapeutic compounds in cucurbits include oxygenated tetracyclic triterpenoids called cucurbitacins, saponins (e.g. cucurbitocitrin in watermelon seeds, tubeimoside 1 in pseudo-fritillary), other glycosides (citrullol and colocynth in *Citrullus colocynthis*), alkaloids (momordicin in bitter melon), ribosome-inactivating proteins (luffaculin in *Luffa operculata*, trichosanthin in *Trichosanthes*), free amino acids (cucurbitin in squash), xanthophylls (lutein in *Cucurbita maxima*) and various other compounds. One interesting glycoside is mogrol I-IV; found in fruits of luo-han-guo, it is 150 times sweeter than sucrose and is being investigated as a non-calorific sugar substitute.

The most intensely studied of these biodynamic compounds are the bitter cucurbitacins, because of their unwanted presence in fruits of the major cucurbit crops. Cucurbitacins in cultivated cucurbits can be a very serious problem, not just due to the extremely unpleasant flavour they impart, but also because of health concerns. These compounds are potent purgatives and laxatives, and in high concentrations can cause serious illness and death.

Cucurbitacins commonly occur in the *Cucurbitaceae*, hence their name, as well as in a few species of other families. There are many different cucurbitacin compounds, existing as glycosides or free aglycones. Those in squash are particularly toxic. However, the highest concentrations (>1%) are found in the fruits of colocynth and various wild species of *Cucumis*. Within a particular plant, concentrations are usually highest in the fruits and roots, and lower in leaves, stems and growing tips, although concentrations may vary in a plant part during development.

Different letters of the alphabet have been assigned to different cucurbitacin compounds. These distinct, but structurally related compounds have some chemotaxonomic value, since some species and groups of species can be characterized by the cucurbitacins they contain. Thus, while only cucumber has cucurbitacin C, squash species are characterized by B, D, E, I and the glycoside of E.

The bitter, toxic cucurbitacins were of selective value during cucurbit evolution, by deterring insects and other herbivores from consuming foliage and fruits. Aphids, spider mites and other pests are repelled by cucurbitacins, but cucumber beetles are not. During the coevolution of plants and insects, cucumber beetles developed an attraction for plants with high concentrations of cucurbitacins.

Wild species of *Cucurbita* have very bitter fruits, and a key step in squash domestication was selection for types with non-bitter fruits. The first plant part consumed by people was probably the seed, which is non-bitter even when fruit flesh is bitter. The cucurbitacins are concentrated in the placenta, the tissue attached to the seed, and the seeds are not bitter if thoroughly cleaned and freed of placenta and adhering flesh. During the process of extracting seeds, bitterness from the flesh would have transferred to the fingers of early cultivators; consequently, mutants with non-bitter flesh would probably have been noticed and their seeds saved for planting. In this way, non-bitter squash was probably selected for and cultivated by native Americans thousands of years ago.

Cucurbitacins occur in the fruits of all squash cultivars, but normally they are in such low concentrations that they cannot be tasted. Occasionally, however, relatively high concentrations make a squash quite bitter. There have been a number of cases where people became very ill, requiring hospitalization after eating only a few grams of bitter-tasting fruit.

Bitterness in cucurbits is influenced by both genetic and environmental factors. At least five genes regulate cucurbitacin biosynthesis, with organ-specific genes controlling the quality and quantity of cucurbitacins in different plant parts. In squash, problems encountered with bitterness are believed to be primarily genetic in origin. Many ornamental gourds, as well as wild populations of C. pepo, have a single dominant allele for bitter fruit. This allele can be transferred to C. pepo squash cultivars by pollinating insects, and plants of future generations may then have bitter fruits. There is no xenia for fruit bitterness; if squash crosses with a gourd or another source of the allele for bitter fruit, the bitterness is not expressed until the F_1 generation.

Although bitterness is most frequently a problem with C. pepo cultivars, probably as a result of gene flow from bitter cultivars or wild populations, it has also been reported in the other squash species. As a result of gene interaction, bitter fruits can develop in the progeny of C. pepo × C. argyrosperma, even when both parents have non-bitter fruits.

In cucumber, environmental factors often influence when bitterness occurs. One fruit may be bitter, while another fruit on the same plant, developing under different weather conditions, may be non-bitter. Some cucumber varieties have bitter foliage but never have bitter fruits; others have bitter foliage and may have bitter or non-bitter fruits, depending on environmental conditions. It is difficult to distinguish between these types in a breeding programme because of the environmentally-induced variability. However, Dutch breeders developed a better method of breeding cucumbers for non-bitter fruits. Andeweg and Bruyn (1958) reasoned that if they could find a cucumber mutant with non-bitter foliage, it would have non-bitter fruits regardless of the environment and growing conditions. After tasting the cotyledons of 15,000 cucumber seedlings, these researchers found one that was non-bitter. Upon growing the seedling out, it produced non-bitter fruits under all growing conditions. The single recessive allele involved has been bred into many modern cultivars.

Moisture and Solids Content in Fruits

Fruits of most cucurbits have a high moisture content (Table 2.2). Inhabitants of the Kalahari Desert in Africa and some other arid areas consume watermelons as a source of water. Cucumbers have 95% moisture content, higher than that for most other vegetables, and they have little nutritional value. Wax gourd fruits also are high in moisture content and low in nutrients. In contrast, baked winter squash has less moisture content

Table 2.2. Nutrient composition per 100 g of fresh cucurbits.[1]

Crop	Moisture (%)	Calories (kcal)	Protein (g)	Fats (g)	Carbo-hydrates (g)	Fibre (g)	Minerals Ca (mg)	P (mg)	Na (mg)	Mg (mg)	K (mg)	Fe (mg)	Zn (mg)	Cu (mg)	Vitamins A (IU)	C (mg)	Thiamin (mg)	Riboflavin (mg)	Niacin (mg)	Pantothenic acid (mg)	B-6 (mg)	Foliacin (Folic acid) (mcg)	Biotin (mcg)
Bottle gourd, fruit	92	26	0.7	0.2	6.3	1.5	18	21	–	–	–	0.5	–	–	0	19	0.01	0.00	0.17	–	–	–	–
Chayote, fruit	92	28	0.6	0.1	7.1	0.7	13	26	5	14.0	102	0.5	–	–	20	19	–	–	–	0.48	–	–	–
root	79	79	2.0	0.2	17.8	0.4	7	34	–	–	–	0.8	–	–	–	19	0.05	0.03	0.90	–	–	–	0.4
Cucumber, whole	95	15	0.9	0.1	3.4	0.6	25	27	6	12.0	160	1.1	0.12	0.01	250	11	0.03	0.04	0.20	0.25	0.04	(16)	0.4
peeled	96	14	0.6	0.1	3.2	0.3	17	18	6	10.0	160	0.3	–	–	–	11	0.03	0.04	0.20	–	0.05	15	–
Melon, 'Casaba'	92	32	1.2	0.1	6.5	0.5	14	16	12	8.0	251	0.4	–	–	30	13	0.02	0.04	0.40	–	–	–	–
Cantaloupe	91	30	0.6	0.1	7.5	0.3	14	16	12	8.4	251	0.4	0.14	0.01	3400	33	0.05	0.03	0.60	0.25	0.06	30	3.0
'Honey Dew'	91	33	0.8	0.3	7.7	0.6	14	16	12	6.7	251	0.1	0.00	0.04	40	23	0.04	0.03	0.60	0.21	0.06	–	–
Pumpkin, fruit	92	26	1.0	0.1	6.5	1.1	3	10	12	12.0	340	0.8	–	0.10	1600	9	0.05	0.11	0.60	–	–	–	–
seed	4	553	29.0	46.7	15.0	1.9	51	1744	1	–	–	11.2	–	–	70	–	0.24	0.19	2.40	–	0.09	36	–
flower	95	16	1.4	0.3	2.7	0.6	47	86	–	–	–	1.0	–	–	200	18	0.02	0.11	0.60	–	–	–	–
Smooth loofah, fruit	95	18	0.8	0.2	4.1	0.5	19	33	–	–	–	0.9	–	–	380	8	0.03	0.04	0.40	–	–	–	–
Summer squash																							
green cultivars	95	17	1.2	0.1	3.6	0.6	28	29	1	23.7	202	0.4	–	–	320	19	0.05	0.09	1.00	0.36	0.08	–	–
'Scallop'	93	21	0.9	0.1	5.1	0.6	28	29	1	23.0	202	0.4	–	–	190	18	0.05	0.09	1.00	0.36	0.08	31	–
yellow cultivars	94	20	1.2	0.2	4.3	0.6	28	29	1	22.0	202	0.4	–	–	460	25	0.05	0.09	1.00	0.36	0.08	31	–
Winter squash																							
'Acorn'	86	44	1.5	0.1	11.2	1.4	31	23	1	32.0	384	0.9	0.14	0.12	1200	14	0.05	0.11	0.60	0.40	0.15	17	–
'Butternut'	84	54	1.4	0.1	14.0	1.4	32	58	1	16.5	487	0.8	–	–	5700	9	0.05	0.11	0.60	0.40	0.15	12	–
'Hubbard'	88	39	1.4	0.3	9.4	1.4	19	31	1	19.0	217	0.6	–	–	4300	11	0.05	0.11	0.60	0.40	0.15	12	–
Watermelon, fruit	93	26	0.5	0.2	6.4	0.3	7	10	1	10.2	100	0.5	0.09	0.02	590	7	0.03	0.03	0.20	0.30	0.07	8	3.6
seed	5	536	22.7	41.2	27.5	2.5	82	483	6	–	606	7.7	–	–	17	–	0.22	0.10	2.60	–	–	–	–
Wax gourd, fruit	96	13	0.4	0.2	3.0	0.5	19	19	6	–	111	0.4	–	–	0	13	0.04	0.11	0.40	–	–	–	–

[1]Modified from Foods and Nutrition Encyclopedia, by Audrey H. Ensminger et al. (1983) with the permission of Pegus Press, 648 West Sierra Avenue, Clovis, California, 93612, USA.

Table 2.3. Nutrient composition per 100 g of processed cucurbit fruits.[1]

Crop	Moisture (%)	Calories (kcal)	Protein (g)	Fats (g)	Carbo-hydrates (g)	Fibre (g)	Minerals								Vitamins							
							Ca (mg)	P (mg)	Na (mg)	Mg (mg)	K (mg)	Fe (mg)	Zn (mg)	Cu (mg)	A (IU)	C (mg)	Thiamin (mg)	Riboflavin (mg)	Niacin (mg)	Pantothenic acid (mg)	B-6 (mg)	Foliacin (Folic acid) (mcg)
Cucumber, pickled																						
cucumber, dill	93	11	0.7	0.2	2.2	0.5	26	21	1428	12.0	200	1.0	0.27	–	100	6	–	0.02	0.07	–	0.01	1.0
cucumber, fresh	79	73	0.9	0.2	17.9	0.5	32	27	673	–	–	1.8	–	–	140	9	–	0.03	–	–	0.01	–
cucumber, sour	95	10	0.5	0.2	2.0	0.5	17	15	1353	–	–	3.2	–	–	100	7	–	0.02	–	–	0.01	–
cucumber, sweet	61	146	0.7	0.4	36.5	0.5	12	16	–	–	–	1.2	–	–	90	6	0.00	0.02	–	–	0.01	–
relish, sour	93	19	0.7	0.9	2.7	1.1	29	20	712	–	–	1.1	–	–	–	–	–	–	–	–	–	–
relish, sweet	63	138	0.5	0.6	34.0	0.5	20	14	–	–	–	0.8	–	–	–	–	–	–	–	–	–	–
Melon, cantaloupe, and 'Honeydew', frozen in syrup	83	62	0.6	0.1	15.7	0.3	10	12	9	13.1	188	0.3	0.13	0.03	1540	16	0.03	0.02	0.50	0.22	0.05	(29.6)
Pumpkin, canned	90	33	1.0	0.3	7.9	1.3	25	26	2	18.0	240	0.4	0.19	0.05	6400	5	0.03	0.05	0.60	0.40	0.04	15.0
Summer squash																						
green cultivars, boiled	96	12	1.0	0.1	2.5	0.6	25	25	1	–	141	0.4	0.18	–	300	9	0.05	0.08	0.80	–	–	–
'Scallop', boiled	95	16	0.7	0.1	3.8	0.6	25	25	1	–	141	0.4	–	–	180	8	0.05	0.08	0.80	–	–	–
yellow cultivars, boiled	93	21	1.4	0.1	4.7	0.6	14	32	3	16.0	167	0.7	0.30	0.14	140	8	0.06	0.04	0.40	–	–	–
Wax gourd, sugared	20	285	0.2	0.2	79.2	0.2	93	17	–	–	–	3.4	–	–	0	0	0.00	0.00	0.00	–	–	–
Winter squash																						
'Acorn', baked	83	55	1.9	0.1	14.0	1.8	39	29	1	–	480	1.1	0.24	0.08	1400	13	0.05	0.13	0.70	–	–	–
'Acorn', boiled	90	34	1.2	0.1	8.4	1.4	28	20	1	–	269	0.8	–	–	1100	8	0.04	0.10	0.40	–	–	–
'Butternut', baked	80	68	1.8	0.1	17.5	1.8	40	72	1	–	609	1.0	–	–	6400	8	0.05	0.13	0.70	–	–	–
'Butternut', boiled	88	41	1.1	0.1	10.4	1.4	29	49	1	–	341	0.7	–	–	5400	5	0.04	0.10	0.40	–	–	–
'Hubbard', baked	85	50	1.8	0.4	11.7	1.8	24	39	1	–	271	0.8	–	–	4800	10	0.05	0.13	0.70	–	–	–
'Hubbard', boiled	91	30	1.1	0.3	6.9	1.4	17	26	1	–	152	0.5	–	–	4100	6	0.04	0.10	0.40	–	–	–

[1] Modified from *Foods and Nutrition Encyclopedia*, by Audrey H. Ensminger *et al.* (1983) with the permission of Pegus Press, 648 West Sierra Avenue, Clovis, California, 93612, USA.

(80–85%) than most other cooked vegetables (Table 2.3).

Carbohydrates entering developing fruits from leaves are transformed by fruit enzymes into soluble sugars. Reducing sugars (e.g. glucose and fructose) are primarily produced in young fruits, with the percentage of sucrose produced increasing with fruit maturity (Fig. 2.5). Sugars are important quality components for squash, melon, watermelon and many other cucurbit fruits. Their content can easily be determined with a hand refractometer. 'Honeydew' melons generally have a higher sugar content (10–16%) than Cantalupensis Group melons (8–14%). Starch conversion, which accounts for greater sugar content in older squashes, continues in mature fruits under storage. In African horned cucumber, sugar content increases during fruit storage, but is even higher if fruits are allowed to ripen on the vine, peaking at 50 days after pollination (Benzioni *et al.*, 1993). During cucumber development, sugar content (mostly glucose and fructose) follows a sigmoidal curve; increase in the parthenocarpic cultivar 'Brilliant' was rapid for the first 6 days following flowering and then slowed until a maximum was reached near the beginning of senescence (Davies and Kempton, 1976).

Among squash species, *Cucurbita maxima* and *C. moschata* produce fruits

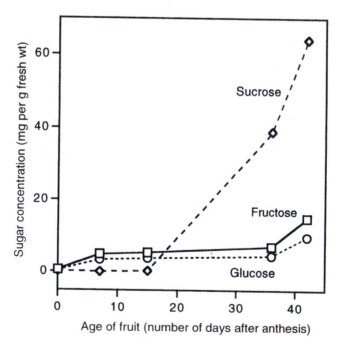

Fig. 2.5. Concentrations of the primary sugars in developing fruits of melon. (Data from Hughes and Yamaguchi, 1983.)

with the strongest taste, highest solids and deepest flesh colour; therefore, they are preferred for commercial canning. Consistency, an important consideration for canning, refers to the stiffness of the processed flesh, which is greatly influenced by starch and soluble solids content. Because the conversion of starch to sugars increases during fruit storage, freshly harvested fruits are chosen for processing.

Composition of pickled cucumbers varies according to the processing method (Table 2.3). Dill pickles and sour pickles are low in caloric content, whereas sweet pickles and relishes have more calories due to the sugar added during processing. Pickles brined during fermentation have a high sodium content.

Nutrients

Cucurbit fruits generally are not very nutritious except for the high vitamin A content of some cultivars of squash and melon and the abundant oil and protein in seeds of *Telfairia, Cucurbita* and other cucurbits. Nevertheless, cucurbits play a significant role in human nutrition, especially in tropical countries where their consumption is high.

On the basis of nutrients produced per man hour invested in production, MacGillivray *et al.* (1942) ranked *Cucurbita maxima* fruit first among all vegetable crops surveyed. This squash had the greatest amount of vitamin A produced per kilogram, per hectare and per man hour. In many nutrient comparisons, however, cucurbits were ranked low when compared with other vegetable crops. Cucumbers were ranked lowest in calories and thiamine, and watermelon lowest in protein and riboflavin per kilogram.

Cucurbit fruits with orange or yellow flesh generally have high concentrations of carotenes, some of which are the precursors of vitamin A. In breeding programmes, it is ideal to have chemical analyses for vitamins and nutrients; but if this is not possible, the breeder can still make significant improvement in carotenoid content simply by selecting for deep orange flesh colour. Usually these carotenoids are the desired provitamin A carotenes (e.g. β-carotene), but in some squash cultivars, the dominant pigments are non-provitamin A xanthophylls like lutein and zeaxanthin. The red and orange coloration in watermelon cultivars is due to the non-provitamin carotenes lycopene and prolycopene, which are closely related to β-carotene but have no known nutritional value. Xishuangbanna gourd from China has orange flesh and more β-carotene than other cucumber cultivars. Navazio (1994), who studied carotenoid content and inheritance in these cucumbers, found that carotene content increases with fruit age, is affected by growing conditions and is controlled by at least two genes. Orange-fleshed melons and squashes are much better sources of β-carotene than cucumbers. 'Honey Dew', 'Casaba' and other green- or white-fleshed melons have lower carotene compositions than melon cultivars with orange flesh.

Leaves, stems and growing tips of cucurbits are eaten in some tropical countries. Cucurbit greens generally contain more calcium, phosphorus, ascorbic acid and iron than the fruits, and can be good sources of vitamin A also. Leaves of various cultivated species contain up to 4–6% protein, while those of fluted pumpkin contain close to 2% fat, and leaves of bitter melon contain about 60 calories per 100 grams.

The most nutritious part of the cucurbit plant is often thrown away. Seeds discarded elsewhere are eaten in some countries, e.g. watermelon seeds in China and squash seeds in Mexico. Various species, including oyster nut, are grown primarily for their edible seeds. Seeds from squash, pumpkin, melon, watermelon and other mature cucurbit fruits are good sources of protein, calcium, phosphorus, iron, magnesium, arginine, methionine, aspartic acid, glutamic acid, thiamin and niacin (Table 2.2). Decorticated, or shelled, seeds are rich in fats (40–60%) and proteins (30–40%), contain few free sugars and lack starch. As in other oilseed crops, cucurbit seed proteins, which consist of mostly globulins and some albumins, have abundant nitrogen-rich amino acids and are deficient in lysine and sulphur-containing amino acids.

Cucurbit seed oils, which are usually edible and suitable for various industrial purposes, are composed primarily of unsaturated fatty acids. The semi-drying oils of squash, melon, watermelon and loofah seeds contain 40–60% linoleic acid, 20–40% oleic acid, 10–20% palmitic acid and 0–15% stearic acid. Cucumber seeds have about 59% oleic, 22% linoleic, 7% palmitic and 4% stearic acids. With an overall flavour of groundnuts, oil in seeds of white-seeded melon is highest in linolenic acid (65%). Seeds of fluted pumpkin, which taste like almonds, are very rich in an oil composed of 37% oleic, 21% palmitic, 21% stearic and 15% linoleic acids. Breeding efforts with buffalo gourd have produced seeds with linoleic acid content ranging up to 82%; however, environmental growing conditions greatly affected the percentage of unsaturated fats.

Pumpkin flowers and chayote roots also have been nutritionally analysed (Table 2.2), with the latter proving to be a good source of easily digestible starch. Research on buffalo gourd roots revealed a dry weight starch content of 55% and a digestibility of 41% when consumed raw and 95% when cooked. Once the cucurbitacins are removed through filtering, processed buffalo gourd starch compares favourably with corn starch in a variety of food industry applications (Gathman and Bemis, 1990).

3

GENETICS AND BREEDING

CYTOLOGY

Chromosome numbers have been determined for the important cultivated species of the *Cucurbitaceae* (Table 3.1) and for a number of other cucurbits as well. The most common haploid numbers are 11 and 12, with 12 considered to be the primitive base number for the family. Squash species have 20 pairs of chromosomes, more than any other cultivated species in the family, and cucumber has the fewest, with only seven pairs. Theories that melon ($n=12$) evolved from cucumber by chromosome fusion, or that cucumber evolved by fragmentation of one of the chromosomes of a melon haploid, were rejected by Ramachandran and Seshadri (1986). Their cytological evidence led them to conclude that melon and cucumber are not closely related in phylogeny. Nor is it likely that chromosome fragmentation from an ancestor having $n=7$ gave rise to other $n=12$ species of *Cucumis*, as 12 is considered to be primitive for the genus (Singh, 1990).

Chromosome morphology is not well characterized for most cucurbits because the chromosomes are typically small and not easily differentiated from the cytoplasm by usual cytological procedures. However, recent improvements in cytological techniques have brought about a better understanding of the cytology and phylogenetic relationships of many cucurbit crop species (Singh, 1990). Variation has been found in some species for length of different chromosomes, position of the centromere, and occurrence of satellites, and Ramachandran and Seshadri (1986) have determined the karyotypes for cucumber and melon.

Heteromorphic sex chromosomes have been identified in dioecious species of *Coccinia* and *Trichosanthes*, but are lacking in various other dioecious cucurbits (Roy and Saran, 1990). In ivy gourd, plants with two X chromosomes are female, and XY plants are male. The Y chromosome is larger and more heterochromatic than the X chromosome and dominant in sex determination; polyploid plants with three X and only one Y chromosome are androecious.

Table 3.1. Chromosome number for some species of the *Cucurbitaceae*.

Species	Haploid number of chromosomes
Benincasa hispida	12
Citrullus lanatus	11
Coccinia grandis	12
Cucumis anguria	12
Cucumis dipsaceus	12
Cucumis melo	12
Cucumis metuliferus	12
Cucumis sativus	7
Cucurbita species	20
Cyclanthera explodens	16
Cyclanthera pedata	16
Lagenaria siceraria	11
Luffa species	13
Momordica balsamina	11
Momordica charantia	11
Momordica cochinchinensis	14
Momordica dioica	14
Praecitrullus fistulosus	12
Sechium edule	12
Trichosanthes cucumerina	11
Trichosanthes dioica	11

Cucurbita is believed to be an ancient tetraploid genus derived from an ancestor with a base chromosome number of ten. Isozymic evidence has confirmed the polyploid nature of this genus. That polyploidy occurred long ago is indicated by all species of the genus having 20 pairs of chromosomes and by disomic 3:1 gene ratios in segregating generations.

Aside from *Cucurbita* and possibly other genera of the tribe *Cucurbiteae*, polyploidy does not seem to have played an important role in the evolution of cucurbit tribes or genera. However, isolated cases of polyploid species are known in some genera with mostly diploid species, e.g. *Cucumis* and *Trichosanthes*, and polyploid cytotypes occasionally arise in melon and a few other species (e.g. kaksa).

Tetraploids have been induced in cucumber, melon, squash, watermelon, loofah and bottle gourd by treatment with colchicine. They do not appear to have any horticultural value for cucurbit crops, except in the production of seedless triploid watermelons.

Within-species triploids have been created in cucumber, melon, squash and watermelon by crossing tetraploids with diploids. Triploids are highly sterile, whereas tetraploids are more fertile than triploids but less so than diploids. The sterility of triploids, due to embryo abortion, has been utilized to breed watermelons with few seeds (see Chapter 4 for more details). Although the use of triploidy in breeding other cucurbits has been investigated, it has not been adopted in varietal development. Naturally occurring triploids have been found in pointed gourd and ivy gourd.

Haploids, which are highly sterile, occasionally occur spontaneously in cucumber and melon. In cucumber, they can be recognized by their reduced seed weight. Haploids can also be produced by interspecific hybridization and pollination with irradiated pollen followed by embryo culture. Haploids treated with colchicine produce homozygous diploid lines more quickly than inbred lines can be developed by selfing.

Polysomaty, in which the chromosome number of some somatic cells of a plant are multiples of the typical chromosome number for that plant, has been detected in melon, squash and other cucurbits. It is believed that triploid plants of pointed gourd were propagated vegetatively from triploid shoots on diploid plants (Singh, 1990).

In squash, sterile interspecific F_1 hybrids have been treated with colchicine to produce amphidiploids. Self-fertile amphidiploid lines with the parentage of *Cucurbita moschata* × *C. maxima* have been produced which segregated for some horticulturally favourable characteristics. However, amphidiploids have not been important in the development of new cultivars.

Triploid interspecific hybrids of *Cucurbita* have been produced, some combining the genomes of three different species. Interspecific triploids have also been backcrossed with one of the diploid parents to create fertile interspecific trisomics. Interspecific trisomics of *C. moschata* and *C. palmata*, combining 20 chromosomes of the former and one of the latter, were synthesized and used to relate genes to specific chromosomes (Graham and Bemis, 1990).

GENES

The large field space requirements for cucurbits and the need for laborious hand pollinations for selfing and crossing cucurbits have been constraints for genetic investigations. Consequently, the genetic knowledge of cucurbits has lagged behind that of such crops as tomato, corn and peas, despite the considerable natural genetic variation in many species of the *Cucurbitaceae*. Seedling morphology genes and molecular marker genes, which reduce space and time requirements, have expedited genetic research in recent years. The establishment of the Cucurbit Genetics Cooperative (CGC), along with the publication of the *Cucurbit Genetics Cooperative Report* annually since 1978, has fostered communication among cucurbit researchers and stimulated more cytogenetic investigations.

Dominance relationships of melon genes were first investigated by

Sagaret in the mid-nineteenth century. Mendelian inheritance in cucumber was reported in 1913, and many genes for each of the major cucurbit crops have since been identified. A total of 170 individual genes were known for the *Cucurbitaceae* when Robinson *et al.* reviewed the genetics of the family in 1976. Of these, 68, nearly half of the genes known then, were for cucumber, followed by 37 for melon, 30 for squash species, 25 for watermelon and 10 for other genera. Numerous other genetic factors were known for cultivated cucurbits and used in breeding programmes, but were not included in the gene list because their inheritance was complex or unknown.

Many additional cucurbit genes and alleles have been identified in the past two decades. The CGC publishes an updated gene list for one of the major cucurbit crops each year, and the most recently published gene lists include 146 loci for cucumber (Wehner, 1993), 100 for melon (Pitrat, 1994a), 81 for watermelon (Rhodes and Zhang, 1995) and 79 for species of squash (Hutton and Robinson, 1992). Thus, the number of known genes for the major cucurbit crops has nearly doubled in the past 20 years. Quite a few of these newly recognized genes code for isozymes. In addition, many multiple alleles have been identified at some loci, e.g. 87 for watermelon isozyme loci alone. A number of distinctive alleles are of importance in breeding and are described below.

Regarding gene nomenclature, we should point out that most genes (though not those of enzymes) are recognized and named according to the discovery of an atypical expression of that gene, which is itself caused by a newly found allele at a previously unrecognized locus. The gene and atypical, or 'mutant', allele are given the same designation (e.g. *gl* is an allelic form of gene *gl*). The first letter of the symbol is in lower case if the atypical allele is recessive and uppercase if dominant. The designation for the normal allele of that gene is given as +, the gene name with a superscripted +, or the gene name with the first letter in the case (lower or uppercase) opposite that for the atypical allele. Continuing the above example for the glabrous (*gl*) gene, the common or normal allele can be designated as +, gl^+ or *Gl*. In this book, we will use the superscript system for referring to common alleles.

When additional alleles are found and assigned to a gene, they are designated with different superscripts appended to the gene name. However, allelism testing runs far behind the discovery of genetic anomalies. When several alleles affect the same heritable trait, they are often treated as belonging to different genes until proven otherwise, although allelism tests are recommended before proposing another locus. This currently is a problem for many alleles relating to disease resistance in cucurbits. More testing is needed to assign various alleles to their proper gene locus and to determine gene linkages.

Cucumber

Many alleles conferring disease resistance have been found and incorporated into cucumber cultivars. Dominant alleles have been reported for resistance to bacterial wilt (*Bw*), scab (*Ccu*), target leaf spot (*Cca*), fusarium wilt (*Foc*) and watermelon mosaic virus (*Wmv*). A dominant allele (*Cmv*) at a single locus has also been postulated for resistance to cucumber mosaic virus, but alleles of additional genes are needed for a high level of resistance. Recessive alleles provide resistance to angular leaf spot (*psl*), zucchini yellow fleck virus (*zyf*) and zucchini yellow mosaic virus (*zymv*). Alleles *Ar* and *cla* reportedly confer resistance to different races of anthracnose. Resistance to papaya ringspot virus is provided by *prsv* or *Prsv-2*. Several genes have been proposed to govern resistance to downy mildew and powdery mildew.

The bitterfree allele (*bi*) inhibits biosynthesis of cucurbitacin, which is an attractant for cucumber beetles but a repellant for spider mites, aphids and various other insects. The *bi* gene is epistatic to the bitter gene (*Bt*) for increased cucurbitacin content.

Salinity tolerance is influenced by many genes, in addition to a single major gene (*sa*). A single gene (*Sd*) has also been suggested to control resistance to sulphur dioxide air pollution.

Compact plant habit, due to short internodes, is produced by alleles *bu*, *by*, *cp*, *cp-2* and *dw*. Although these alleles are currently assigned to distinct loci, allelism testing is needed to confirm that none represent allelic variants of the same gene. Allele *dw* also retards the development of oversized, and hence unmarketable, fruits. Vine size is also reduced by the determinate habit allele (*De*), which is modified by the intensifier gene *In-De*.

The interaction of two major genes, *m* and *F*, influences sex expression. *F* is modified by *In-F* and alleles at other genes. Some cultivars heterozygous for *F* have only female flowers, due to the right combination of modifying genes, but others can be monoecious under some growing conditions. Cultivars homozygous for *F* have been developed because they are more dependably gynoecious than heterozygous cultivars. The use of these genetic backgrounds in the production of gynoecious hybrid seed is described in Chapter 6. Additional genes have been reported to influence andromonoecious (*m-2*) and gynoecious (*gy*) sex expression.

Multipistillate alleles *mp* and *Mp-2* and their modifiers at other loci increase the number of pistillate flowers per node. The development of parthenocarpic fruits is governed by *Pc* and modifying genes.

Fruit spine colour is governed by the putative genes *B*, *B-2*, *B-3* and *B-4*, with black spines being dominant to white. Spine number and size is influenced by genes *s-1*, *s-2* and *s-3*. Spines and warts are absent on fruits of plants with the *gl* allele for glabrous foliage, and more pronounced when possessing the tuberculate fruit allele, *Tu*.

European glasshouse cultivars have glossy fruits with a tender skin and uniform dark green colour, without light green stippling. These are monogenic traits, governed by the D gene for dull versus glossy fruit and the tender skin (te) and uniform colour (u) genes.

Green immature fruit colour is dominant to white (w) and yellow green (yg). The interaction of alleles at two genes, wf and yf, reportedly determines white versus yellow or orange flesh colour.

Six linkage groups have been proposed for cucumber genes (Pierce and Wehner, 1990), but rigorous testing and recombination frequencies are generally lacking. Chromosome assignments of the linkage groups are unknown.

Melon

Many loci governing disease resistance of melon have been investigated. A single gene, Ac, governs alternaria leaf blight resistance in melon line MR-1. Resistance to races 0 and 2 of fusarium wilt is provided by $Fom-1$ or $Fom-3$, and there is an allele of another gene ($Fom-2$) for resistance to races 0 and 1. Allele Mc confers a high level of resistance to gummy stem blight, and $Mc-2$ a moderate degree of resistance to that disease. The nsv gene governs resistance to melon necrotic spot virus. Eight loci have been postulated to influence resistance to the powdery mildew disease caused by $Sphaerotheca$ $fuliginea$, and three more to the powdery mildew incited by $Erysiphe$ $cichoracearum$. Four genes have been reported for downy mildew resistance. Two alleles (Prv^1 and Prv^2) of a single gene provide resistance to papaya ringspot virus, but they differ in reaction to some strains of that virus. Prv^2 is recessive to Prv^1 but dominant to Prv^+. Resistance to pathotype 0 of zucchini yellow mosaic virus is provided by dominant allele Zym.

Several melon genes for insect resistance have been identified. Af influences resistance to red pumpkin beetle. Tolerance to melon aphid is provided by the dominant allele at gene Ag, and Vat conditions resistance to viruses transmitted by that pest. Two complementary genes, $dc-1$ and $dc-2$, govern resistance to melon fruit fly. Genes controlling foliar cucurbitacin content, including cb and Bi, influence insect resistance; plants homozygous recessive at both of these loci are resistant to cucumber beetles.

Many genes influence melon plant habit, several of which have been identified. Compact habit can be obtained by breeding for the homozygous recessive state at one of the short internode loci, $si-1$, $si-2$ or $si-3$. Allele Imi increases internode length on the main stem, and ab (abrachiate) inhibits lateral branch development.

Genes a (andromonoecious) and g (gynomonoecious) interact to influence sex expression in the following manner: monoecy ($a^+/-$ $g^+/-$), andromonoecy (a/a $g^+/-$), gynomonoecy ($a^+/-$ g/g) and hermaphrodism (a/a g/g). Stable gynoecious sex expression can be achieved by combining homozygous recessive *gy* (gynoecious) at a third locus with a dominant allele at the *a* locus and the *g* allele in the homozygous state at the *g* locus (i.e. $a^+/-$ *g/g gy/gy*). Melon plants are male sterile if homozygous recessive for alleles at one of the independent genes *ms-1*, *ms-2*, *ms-3*, *ms-4* or *ms-5*.

Fruits of some melon cultivars detach from the vine at maturity, but other cultivars have persistent fruits. Two dominant alleles, *Al-1* and *Al-2*, control formation of the abscission layer.

Fruit quality is a polygenic trait, but several individual genes have a major effect. In wild melon populations, fruits may be bitter due to *Bif* and have mealy flesh texture because of the *Me* allele. Sour taste is dominant to sweet and conditioned by the *So* gene. A recessive allele (*jf*) for juicy flesh and a dominant allele (*Mu*) for musky flavour have been reported.

Many genes influence the intensity of flesh colour, but individual major genes may determine whether the flesh is orange (which is dominant), green or white. Recessive alleles of two genes govern green flesh (*gf*) and white flesh (*wf*), with wf^+ epistatic to gf^+/gf.

External fruit colour is influenced by *Mt* (mottled rind), *st* (striped epicarp), *w* (white mature fruit), *Wi* (white immature fruit) and *Y* (yellow epicarp). Fruit shape genes include *O* (oval), *s* ('sutures' or vein tracts) and *sp* (spherical fruit shape).

A total of 13 linkage groups have been described for melon, but not all are independent since there are only 12 chromosomes, and their chromosome locations are unknown (Pitrat, 1994b).

Squash

Fewer disease resistance genes have been reported for squash than for the other major cucurbit crops. Powdery mildew resistance of *Cucurbita okeechobeensis* (Small) L. H. Bailey and *C. lundelliana* L. H. Bailey is controlled by a single dominant allele, *Pm*, and modifier genes. A single gene, *Zym*, has been reported to affect resistance to zucchini yellow mosaic virus.

Resistance to the melon fruit fly is provided by allele *Fr*. *Cucurbita pepo* plants homozygous for *cu*, which reduces foliar cucurbitacin content, are not preferred by cucumber beetles since these insects are attracted by cucurbitacins. The *Bt* allele of another cucurbitacin gene is responsible for very bitter fruits.

Resistance in *C. moschata* to the herbicide trifluralin is governed by gene *T*, which is modified by an inhibitor gene, *I-T*.

A single gene (*G*) governing gynoecious sex expression occurs in *C. foetidissima*, but the dominant allele has not been transferred to the cultivated squash species. Two male-sterile genes, *ms-1* and *ms-2*, are known for *C. pepo*.

Bush habit due to short internodes is conditioned by allele *Bu* in *C. pepo*; expression of the allele is as a dominant in early stages of plant development and as a recessive later on. Bush habit of *C. maxima* is also under monogenic control.

Fruit colour of different squash cultivars is quite diverse, and a number of fruit pigment genes have been investigated. The incompletely dominant allele of the *B* gene, which was found in an ornamental gourd, has been used to breed *C. pepo* cultivars with golden fruit colour and high vitamin A content. This allele affects many other horticultural traits, from thinner fruit shape to reduced storability and increased sensitivity to chilling injury. The fruit colour expression of the *B* allele is influenced by pigmentation extender genes *Ep-1* and *Ep-2*, suppressor gene *Ses-B* and other modifying genes. Yellow to orange fruit colour of *C. maxima* cultivars is controlled by the *B-2* gene and modifiers. Two genes of *C. pepo*, *l-1* and *l-2*, govern light pigmentation of fruit. Allele *l-1*St (striped fruit) is recessive to allele *l-1*$^{+}$ but dominant to *l-1*. Two genes, *r* and *W*, have been reported to govern white rind colour in *C. pepo*. The *Wf* allele also controls white rind colour and is responsible for white flesh, whereas the *Y* allele confers yellow fruit colour in *C. pepo*. Green fruit colour in *C. moschata* is governed by gene *Gr*. The *Mldg* gene of this species determines whether immature fruits are mottled light and dark green or coloured uniformly dark green. In *C. maxima*, blue fruit skin colour is produced by allele *bl* and red by *Rd*.

Wild *Cucurbita* species and some summer squash cultivars have a hard fruit rind. Rind hardness is influenced by the *Hr* (hard rind) gene of *C. pepo* and *Hi* (hard rind inhibitor) of *C. maxima*.

Disc fruit shape is governed by the *Di* gene of *C. pepo*, and warty fruit surface is conditioned by the *Wt* gene. The characteristic post-cooking stringiness of the flesh of 'Vegetable Spaghetti' (*C. pepo*) is reportedly governed by a single gene, *fl*. A recessive allele of a major gene (*n*) and alleles at modifying genes for naked seed inhibit lignification of the seed coat in *C. pepo*.

Based on the cross of *C. maxima* × *C. ecuadorensis* Cutler & Whitaker, five linkage groups of isozyme genes were proposed by Weeden and Robinson (1986).

Watermelon

Two genes of watermelon have been reported to govern anthracnose resistance, *Ar-1* for race 1 and *Ar-2* for race 2 resistance. Alleles *db* and *Fo-1* provide resistance to gummy stem blight and race 1 of fusarium wilt, respectively. Susceptibility to powdery mildew is governed by *pm*. A single dominant allele, *Zym*, confers resistance to zucchini yellow mosaic virus. Insect resistance genes for watermelon include *Af* (red pumpkin beetle resistance) and *Fwr* (fruit fly resistance).

Watermelon plants with short vines can be bred if they are homozygous recessive at the *dw-2* locus or for one of the two alleles for dwarf plant habit at the *dw-1* locus. Branching at the lower nodes of the main stem is reduced by allele *bl*.

Andromonoecy is recessive to monoecy and conditioned by gene *a*. Two alleles are known which produce male sterility, *ms* and *gms*, the latter associated with glabrous foliage.

When plants are homozygous recessive for the *e* (explosive rind) allele, the fruit is thin and tender, bursting when cut. The *f* gene determines furrowed fruit surface. The incompletely dominant allele *O* governs elongate versus spherical fruit shape.

Dark green skin colour (g^+) is dominant, and two other alleles at the *g* locus determine light green (*g*) or striped green skin (gs^s). Greenish mottling of the exocarp is produced by the *m* allele, and pencilled lines by *p*. Golden mature fruit colour and chlorosis of older leaves is governed by gene *go*.

A gene with a dominant allele for white flesh (*Wf*) is epistatic to a gene for yellow flesh; the double recessive is red-fleshed. Allele *C* produces canary yellow flesh colour. Yellow flesh due to allele *y* is recessive to red flesh (y^+); another allele, y^o (orange flesh), is recessive to y^+ but dominant to *y*. The *su* allele suppresses fruit bitterness.

The interaction of alleles at several genes, including *d* (dotted seed coat). *r* (red), *t* (tan) and *w* (white seed coat), determine seed coat colour and pattern. Seed size is influenced by the *l* (long) and *s* (short seed) genes and a third gene, *X*. Five genes for seed protein composition (*Spr-1*, *Spr-2*, *Spr-3*, *Spr-4* and *Spr-5*) are included in the watermelon gene list.

Four linkage groups of 13 isozyme genes were reported by Navot and Zamir (1986).

THE MECHANICS OF PLANT BREEDING

Manual Pollination

Large unisexual flowers and sticky pollen facilitate manual pollination in cucurbit breeding experiments. In squash, male and female flowers to be used the next morning are chosen the previous day. The closed corolla of an appropriate squash flower will have a light touch of yellow at the apex. The corolla tips of the staminate flower may be slightly separated also (Fig. 3.1). The apices of the chosen flowers are taped or tied to prevent pollinator entry. The next morning, a staminate flower is removed, opened, and the dehisced pollen rubbed on to the stigma of a previously tied female flower. After pollination, the female flower is bagged or the corolla resecured to prevent pollinator contamination, and the developing fruit is then tagged.

Because squash pollen does not store well, the use of fresh pollen is customary in squash breeding. However, pollen can be used from pre-anthesis flowers that are kept for a few days at low temperature and high humidity.

Fig. 3.1. Male and female squash flowers one day before anthesis, at which time, the corollas are tied shut to exclude insects from flowers that are to be hand-pollinated. (Whitaker and Robinson, 1986. Reprinted courtesy of Chapman & Hall, New York.)

Hand cross-pollination of andromonoecious melons is more difficult than for monoecious cultivars since the anthers need to be manually excised from each perfect flower. The anthers of perfect flowers are usually removed on the afternoon before anthesis; afterwards, the emasculated flower is enclosed to prevent insect pollination. As with squash. the male melon flowers are either closed before anthesis or picked and kept indoors overnight at high humidity; pollen transfer from these flowers takes place the following morning.

Tissue Culture

Tissue culture techniques in cucurbits are used primarily for embryo rescue, propagation, and embryogenesis and organogenesis. Success in producing interspecific plantlets by embryo culture has been achieved in *Cucurbita* (*C. pepo* × *C. moschata*, *C. pepo* × *C. ecuadorensis*) and *Cucumis* (*C. metuliferus* × *C. zeyheri* Sond.). Researchers are still struggling to produce plants from rescued embryos from the cross of African horned cucumber and melon. So far, there are no reports of the successful use of protoplast fusion in creating interspecific hybrids.

Cultivars within a cucurbit crop often differ in their ability to be regenerated through tissue culture. Embryos have been obtained from various plant tissues (e.g. leaf callus, hypocotyl explants, protoplast-derived callus), but their development into plants has been sporadic. Cotyledons and hypocotyls have proven the best explant tissues for organogenesis, although leaves, fruit tissues and embryos have been used also.

Genetic engineering has been a boon to the refinement of methods for regenerating cotyledon explants. For example, in their production of transgenic melon plants, Gonsalves *et al.* (1994) cultured the developing tissue on three different variations of the Murashige and Skoog medium according to plant developmental stage (i.e. regeneration, shoot elongation and rooting stages). They also performed regeneration experiments in which they found that proximal explants produced a higher percentage of regenerated shoots than distal tissue samples, with most shoots arising from the proximal side of the square-cut piece of cotyledon.

Achieving Breeding Objectives

In addition to genetic engineering (this chapter, p. 55), current trends in cucurbit breeding include the use of multivariate statistical analyses (e.g. to increase selection efficiency by evaluating the genetics behind heterosis), research on gene mapping and linkage, the induction of mutations with chemicals and radiation, the use of wider crosses within species and gene transfer among species, and field screening of germplasm in genebanks for desirable characteristics.

Although genetic engineering and backcrossing typically focus on single alleles, more attention is being given to multiple gene selections for single or multiple characteristics. For example, in Saudi Arabia, the melon cultivars 'Najd I' and 'Najd II' were bred for tolerance of both high temperature and salinity. In some cases where multiple insect resistance has been found, it is possible to breed for resistance to several insects from a single cross. 'Butternut' squash (*Cucurbita moschata*), for example, is resistant to squash vine borer, squash bug, cucumber beetles, pickleworm, melonworm and leafminers. Sources of combined insect and disease resistance are also known for cucumber and melon. Sometimes, more complex inherited resistance to a disease provides a higher level of resistance to a particular strain as well as protection against more strains of the pathogen than does monogenic resistance. Therefore, a combination of biotechnology and conventional breeding techniques is likely to be important in the future genetic improvement of cucurbits. Already, genetically engineered virus resistance is being combined with conventionally bred resistance to powdery mildew in squash and fusarium wilt in melon (Quemada and Groff, 1995).

Selection for many morphological characteristics is relatively straightforward. Progeny are visually inspected for the desired character state, and selection and further breeding proceed according to objectives. Simple Mendelian genetics often reveal relationships (e.g. dominant, codominant, recessive, complementary, etc.) among alleles at one or two major loci.

Although biochemical qualities can be evaluated with the help of laboratory equipment and analysis (e.g. using a refractometer to select for high sugar content in fruits), apparent linkage or pleiotropy for morphological characters has also been used to breed for these qualities. For example, selection for fruit flesh carotenes initially was based on fruit colour, i.e. orange-fleshed fruits were presumed to have higher concentrations of these organic compounds. However, subsequent research in squash (*C. pepo*) revealed that fruit colour is controlled by the complex interaction of alleles at several genes, and although total carotenoid content is affected by these genes, there are additional genes influencing the percentage of carotenes in the total carotenoid content (Paris, 1994). Also, the major allele (*B*) for high carotenoid content has various pleiotropic, and not always favourable, effects on fruits and foliage. Consequently, breeding for increased carotene content has not been straightforward and continues today.

Gene linkage, or more likely pleiotropy, has also been a problem in breeding monoecious melons. The use of monoecious melons in F_1 hybrid seed production is desirable because emasculation of the female parent is easier with the pistillate flowers of monoecious melons than with the perfect flowers of andromonoecious cultivars. However, selection for monoecy has been limited by its genetic association with elongated fruit shape. This fruit shape is dominant, and F_1 hybrids with a monoecious parent generally have undesirably long fruits. Modifying genes can influence the shape of melons with the monoecious gene, however, and H.M. Munger of Cornell University has had success in breeding monoecious selections with nearly round fruits.

Many objectives of cucurbit breeding programmes are traits of complex inheritance. Earliness, yield, adaptation to certain environmental conditions, quality and other important performance characteristics are influenced by polygenes of undetermined number and interaction. Consequently, larger progeny populations, rigorous experimental design and multivariate analyses are required to unravel genetic backgrounds and evaluate breeding material for future selections. Pedigree and mass selection have been successfully used to improve cucurbits for these traits. Backcrossing has been used also, especially for disease resistance (Munger *et al.*, 1993).

Several studies with cucumber, melon and squash (reviewed in Whitaker and Davis, 1962) indicate that, in general, there is little or no inbreeding depression but there is significant heterosis in certain combinations. Heterosis, particularly for earliness, has kindled interest in breeding F_1 hybrid cultivars, which are very important today for cucumber, melon, squash and watermelon. Increasingly, wider crosses are being employed to produce F_1 populations with more favourable characteristics. The mechanics of F_1 seed production are described in Chapter 6.

Interspecific Hybridization

Sagaret and Naudin tried to cross melon with cucumber in the mid-nineteenth century, but without success. All later investigators trying to cross these distantly related species have also failed. However, interspecific hybrids have been produced for other species of *Cucumis* (Fig. 3.2), as well as for *Cucurbita* (Fig. 3.3), *Citrullus*, *Luffa*, *Momordica*, *Trichosanthes* and other cucurbit genera. Only in *Cucurbita* has interspecific gene transfer been utilized successfully for crop improvement.

Although interspecific hybrids are generally the first step in a long breeding programme, F_1 hybrids of *C. maxima* × *C. moschata* are utilized as cultivars of winter squash without further breeding. Both parental species are monoecious, having many more male than female flowers, but the interspecific hybrid is gynoecious or predominantly female in sex expression. The interspecific hybrid is usually very productive if pollination conditions are favourable (e.g. a monoecious cultivar is grown nearby to provide pollen). The phenomenon of a gynoecious hybrid being produced by crossing two monoecious species also occurs in the cross of *C. pepo* × *C. ecuadorensis*.

The cross *C. maxima* × *C. moschata* is normally difficult to make; many pollinations are usually needed for each fruit set, and only a few seeds per fruit are typically produced. Japanese seedsmen, however, have been able to make this cross so successfully that they market interspecific F_1 seed commercially. Presumably, they have found *C. maxima* and *C. moschata* parents that 'nick' well and are more compatible than most members of these species. 'Tetsaka-buto', the first popular interspecific hybrid squash cultivar, is a cross of 'Delicious' (*C. maxima*) × 'Kurokawa No. 2' (*C. moschata*). Seed production for the interspecific hybrid is most prolific when *C. maxima* is the maternal parent.

The cross *C. maxima* × *C. pepo* is difficult but not impossible to make; the F_1 is highly sterile. *Cucurbita pepo* and *C. moschata* can be crossed, but compatibility is influenced greatly by the genotypes of the parents, with some

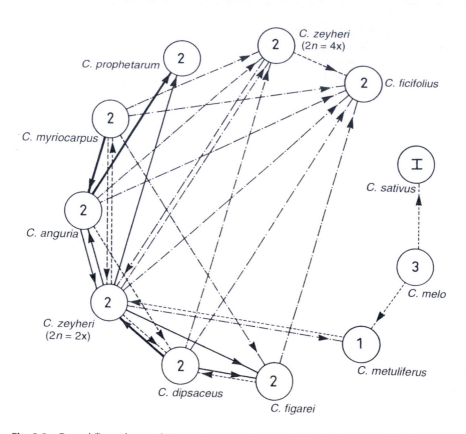

Fig. 3.2. Crossability polygon of *Cucumis* species. Arrows point in the direction of the female parent. Moderately to strongly self-fertile and cross-fertile hybrids (thick solid line); sparingly self-fertile and moderately cross-fertile hybrids (thin solid line); self-sterile, usually not cross-fertile hybrids (dash and dot line); inviable seeds or seedlings (dashed line); absence of a line indicates that seeded fruits were not obtained. (Nijs and Visser, 1985. Reprinted courtesy of Kluwer Academic Publishers.)

cultivars crossing more readily than others. Wall and York (1960) reported that the cross was more easily made when one of the parents was an F_1 hybrid, which increased gametic diversity. The bush gene of *C. pepo* has been introgressed into *C. moschata* to incorporate compact plant habit into that species. Zucchini yellow mosaic virus (ZYMV) resistance, derived from 'Nigerian Local' (*C. moschata*), has been bred into the 'Tigress' and 'Jaguar' cultivars of *C. pepo*.

Wild *Cucurbita* species are being used to breed squash for disease resistance. In disease studies, *C. ecuadorensis* and *C. foetidissima* were found to be resistant to a greater number of viruses than any other species of *Cucurbita* tested (Provvidenti, 1990). Buffalo gourd is very difficult to use in squash

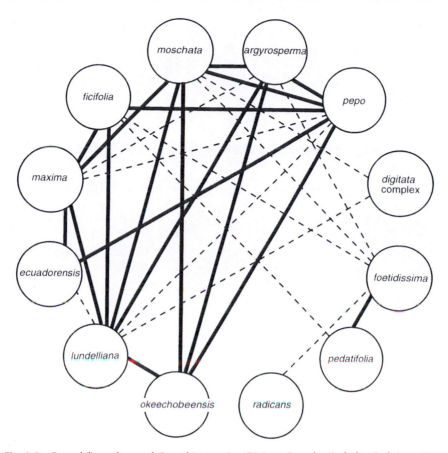

Fig. 3.3. Crossability polygon of *Cucurbita* species. 'Digitata Complex' includes *C. digitata, C. palmata, C. cylindrata* and *C. cordata,* which are considered subspecies of *C. digitata* by some scientists. All crossing combinations have been tried in a least one direction, except for *C. pedatifolia* with *C. maxima,* and *C. radicans* with *C. pedatifolia, C. ficifolia, C. ecuadorensis, C. okeechobeensis,* and *C. digitata sensu lato.* Early works describing hybridization attempts with *C. radicans* were in error as described in Merrick (1990); the material was mis-identified. Other published sources, as well as the unpublished work of Tom Andres (New York, 1996, personal communication), were used to create this diagram. Solid lines indicate a F_1 hybrid that is at least partially fertile; dashed lines indicate a viable but sterile F_1 plant.

breeding because of its distant relationship and incompatibility with the cultivated species (Fig. 3.3). However, virus resistance alleles have been successfully introgressed from *C. ecuadorensis* to *C. maxima.* Multiple virus resistant germplasm derived from *C. maxima* × *C. ecuadorensis* was developed at Cornell University, USA, and provided to seedsmen in 1985. 'Redlands Trailblazer', a winter squash resistant to ZYMV, watermelon mosaic virus (WMV) and papaya ringspot virus (PRSV), was bred in Australia from the

Fig. 3.4. *Cucurbita moschata* (top left) was crossed with *C. okeechobeensis* ssp. *martinezii* (top right). Good horticultural type, combined with disease resistance, was recovered in the BC$_1$ generation (bottom three rows), produced by crossing the F$_1$ (second row) with *C. moschata*.

same interspecific cross. Cultivars of *C. pepo* vary considerably in their compatibility with *C. ecuadorensis* (Robinson and Shail, 1987). This and other distant interspecific *Cucurbita* crosses benefit from embryo culture.

Resistance to cucumber mosaic virus (CMV) and powdery mildew has been transferred from *C. okeechobeensis* to *C. pepo* and *C. moschata*. The cross *C. okeechobeensis* × *C. pepo* is difficult to make, even with embryo culture, and sterility is a serious problem with the interspecific hybrid. These difficulties are overcome by the use of a bridging species, e.g. crossing *C. pepo* to the F$_1$ of *C. okeechobeensis* × *C. moschata* (Whitaker and Robinson, 1986).

Because undesirable traits (e.g. bitterness) of a wild parent are often dominant, unacceptable horticultural types predominate in the F$_2$ generation of these interspecific crosses. However, only a single backcross to 'Butternut' (*C. moschata*) of the F$_1$ of 'Butternut' × *C. okeechobeensis* spp. *martinezii* was needed to produce disease resistant (powdery mildew and CMV) plants with fairly good fruit characteristics (Fig. 3.4). In general, the number of backcrosses to the domesticated parent that are needed depends on the breeding objective and genetics of the parents. Once acceptable progeny are produced, selfing is performed to obtain uniformity.

Wild species of *Cucurbita* could probably be used further in squash breeding programmes to provide resistance to other pathogens and insects. Additional desirable traits, such as the drought tolerance and gynoecious sex expression of *C. foetidissima*, could be transferred as well.

GENETIC ENGINEERING

The nature of DNA in cucurbits has been under study for over ten years. Bhave *et al.* (1986) analysed the distribution of repeat and single copy DNA sequences in smooth and angled loofah, wax gourd and ivy gourd. Around the same time, Ganal and Hemleben (1986) compared restriction enzyme maps of ribosomal DNA repeats in cucumber, melon, and two squash species, *C. maxima* and *C. pepo*. An early and continuing objective for studying DNA and RNA in cucurbits has been to clarify phylogenetic relationships within the family. Analysis of chloroplast DNA variation revealed species relationships in *Cucurbita* (Wilson *et al.*, 1992) and *Cucumis* (Perl-Treves and Galun, 1985), and researchers are beginning to apply these techniques to higher ranks.

Much of the progress made towards understanding the genetic code of cucurbits and other plants is due to the use of restriction enzymes, also called endonucleases. An endonuclease cleaves DNA at a particular spot based on the enzyme's recognition of a certain sequence of nucleotides. There are many different endonucleases, each cutting DNA at different spots, creating DNA fragments of various lengths.

Endonucleases allow for DNA mapping, which is ongoing for cucumber, melon and squash. For example, Gounaris *et al.* (1990) worked to resolve the genetic map of chloroplast DNA in *Cucurbita pepo*, and Reynolds and Smith (1995) sequenced cucumber's isocitrate lyase gene. One use of genetic mapping is for the characterization and identification of new cultivars, which is of concern to those who desire proprietary protection for their cultivar creations. Another use is for detecting linkages between desirable traits, particularly disease resistance, and molecular markers such as restriction fragment length polymorphisms (RFLPs) or random amplified polymorphic DNA (RAPDs). Linkage between a DNA marker and a virus-resistance gene allows one to identify the presence of the resistance gene by checking the genotype at the marker locus. In cucumber, markers are currently being evaluated for linkage to resistance to cucumber mosaic virus (CMV), papaya ringspot virus (PRSV), water melon mosaic virus (WMV) and zucchini yellow mosaic virus (ZYMV).

Endonucleases also enable scientists to create recombinant DNA, which results when cleaved DNA strands from two different DNA molecules are enzymatically joined. If one of these molecules is a bacterial plasmid (an independent, self-replicating ring of DNA in the bacterial cell), then DNA from the other molecule can be incorporated into the plasmid. The plasmid containing the recombinant DNA can be transferred to embryonic plant cells, which, in turn, pass the genetic material on to new cells.

When the transgenic vector is a bacteriophage (a bacteria-infecting virus) instead of a plasmid, the recombinant DNA is passed from virus to bacterium (where it is incorporated into the bacterial DNA), and then from bacterium to plant. Bacteria containing incorporated viral DNA possess a degree of immunity to attacks by the same phage because the incorporated DNA directs synthesis of a phage repressor molecule which regulates the expression of the viral DNA.

In cases where simply inherited sources of disease resistance have not been found (e.g. resistance to CMV in melon and squash), genetic engineering and transfer of the coat protein or replicase genes of the virus to cucurbits has given these cucurbits disease resistance. For example, Chee and Slightom (1991) developed the coat-protein mediated mechanism for CMV protection in cucumber. They used a disarmed strain of *Agrobacterium tumefaciens* to transfer the engineered viral coat protein to cucumber. Small pieces of young, growing cotyledon tissue were soaked in a solution containing the recombinant DNA-carrying bacterial plasmids. The transformed embryogenic calli were regenerated and the resulting plants were inoculated and shown to have resistance to CMV. Gonsalves *et al.* (1992) went on to prove that the transgenic plants had a high level of CMV resistance under natural field conditions, where viral exposure is usually via insect vectors.

Similar field performance experiments have been conducted for transgenic squash lines (*Cucurbita pepo*). The virus-susceptible hybrid 'Pavo' was compared with two transgenic selections from this cultivar that possessed CMV and WMV resistance (Arce-Ochoa *et al.*, 1995). Under field conditions in Texas, the two transgenic selections produced only 3% and 14% symptomatic plants, compared with 53% for 'Pavo'.

In the USA, the first commercially available transgenic cucurbit was 'Freedom II' (previously called 'ZW-20'), a yellow summer squash of the crookneck type (*C. pepo*). This cultivar was engineered with resistance to WMV and ZYMV. Again, *Agrobacterium tumefaciens* was used, this time to transfer the coat protein genes of WMV and ZYMV to the crookneck squash.

Another method of delivering engineered DNA to cotyledon explants uses a helium-driven particle accelerator instead of an agrobacterium. Gonsalves *et al.* (1994) found that they were able to obtain a similar percentage of transgenic melon plants when the plasmid containing the CMV-WL (white leaf strain) coat protein gene was delivered as a microprojectile that penetrated embryonic cells.

For production of a desirable gene product in a host plant, genetic material can be transferred directly from virus to plant via viral infection. α-Trichosanthin is a eukaryotic ribosome-inactivating protein in the cucurbit Chinese snake gourd. Because this protein inhibits replication of HIV *in vitro*, scientists would like to be able to produce large amounts of the protein for further testing. After determining and amplifying the α-trichosanthin coding sequence, the gene, along with a subgenomic promoter controlling the expression of α-trichosanthin, was cloned into a hybrid tobacco mosaic virus. Mature tobacco plants were inoculated with the genetically engineered virus. As the virus spread from inoculated leaves to non-inoculated leaves on a plant, the recombinant α-trichosanthin accumulated in the infected leaves to levels of at least 2% of total soluble protein in less than 2 weeks (Kumagai *et al.*, 1993).

4

MAJOR AND MINOR CROPS

CUCUMBER, MELON AND THEIR RELATIVES

Introduction

A recent, comprehensive biosystematic monograph of the genus *Cucumis* (Kirkbride, 1993) recognizes 32 species, including two major crops, cucumber (*C. sativus*) and melon (*C. melo*), and two minor crops, bur gherkin (*C. anguria*) and African horned cucumber (*C. metuliferus*). Various other species are infrequently cultivated (e.g. *C. dipsaceus*) or collected and utilized for food, water, or medicine (e.g. *C. africanus* L.).

Cucumis is divided into subgenus *Cucumis*, composed of *C. sativus* and *C. hystrix* Chakravarty, and subgenus *Melo* (Mill.) C. Jeffrey, containing the remaining species. Phylogenetic investigations of *Cucumis* species have been based on crossing relationships, karyotypes, flavonoids, isozymes, chloroplast DNA and other criteria. All of these studies agree that the subgenera are widely separated, to the point that it has been proposed they be in distinct genera. The disparity of cucumber originating in Asia, but most other *Cucumis* species being indigenous to Africa, may be a relic of continental drift, separating Africa and Asia and isolating the two subgenera.

Interspecific crossing relationships (Fig. 3.2 and Deakin *et al.*, 1971) suggest four groups of *Cucumis* species: (i) the anguria group, composed of *C. anguria* and seven other intercompatible species (marked '2' on Fig. 3.2) with softly spiny fruits; (ii) *C. metuliferus*; (iii) *C. sativus*, including the fully compatible *C. sativus* var. *hardwickii* (Royle) Gabaev; and (iv) the melo group, which includes *C. melo*, *C. humifructus* and *C. sagittatus* Peyritsch in Wawra & Peyritsch. Species of the melo group have hairy immature fruits, but unlike other *Cucumis* species, their mature fruits lack spines. Interspecific hybrids have not been obtained in this group, although fruits with partly developed seeds were obtained in the cross *C. melo* × *C. sagittatus*.

Puchalski and Robinson (1990) proposed seven groups of *Cucumis* species according to isozyme patterns. The groupings were generally similar to those above, except that the anguria group was separated into three isozyme groups, and *C. sagittatus* was placed in a different isozyme group than *C. melo*.

Minor Crops

Cucumis anguria

Cucumis anguria var. *anguria* is known as bur gherkin and West Indian gherkin. The term gherkin is imprecise, since it has been used both for *C. anguria* and for small fruits of pickling cultivars of cucumber.

Bur gherkin is a fast-growing annual cultivated in Latin America, Brazil, and the West Indies. It is also listed in seed catalogues and occasionally grown by home gardeners in the USA and other countries. In many of these areas, bur gherkin has become a persistent weed.

Cucumis anguria was previously considered to be the only species in the genus native to the Western Hemisphere, and its common name reflects its presumed origin in the Caribbean. Now, scientists believe that this cultigen was introduced to the West Indies and Brazil from Africa by the slave trade. The African taxon once labelled as *C. longipes* Hook. f. and now reclassified as *C. anguria* var. *longaculeatus* Kirkbride is considered the progenitor of domesticated var. *anguria*. A salient step in the domestication of this species was selection for non-bitter fruit, which was accomplished by a single gene mutation.

The flowers, tendrils and lobed leaves of the bur gherkin are smaller than those of cucumber. The oval fruits are light green to pale yellow and covered with short fleshy spines. Although they are smaller (ca 5 cm) than most cucumbers, the peduncle reaches up to about 20 cm in length (Baird and Thieret, 1988), which is much longer than that of cucumber. The plants produce great quantities of fruits throughout the growing season. Immature fruits are eaten fresh, pickled or cooked in curries, soups or stews.

Cucumis dipsaceus

This species is occasionally grown for its ornamental fruit. The small (ca 3–7 cm long), light green fruit is densely covered with fleshy spines measuring about 5 mm long (Fig. 4.1). This species has been called hedgehog gourd and teasel gourd for the resemblance of its fruit to a hedgehog and the spiny flower head of teasel, *Dipsacus fullonum* L., respectively. Although its origin is in northeastern Africa, teasel gourd has escaped from cultivation and become a feral weed in Mexico and other countries.

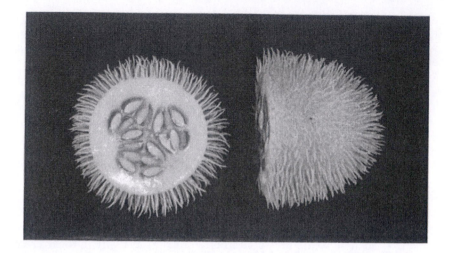

Fig. 4.1. Teasel gourd, measuring ca 4.5 cm in diameter at cross-section, including spines.

Cucumis metuliferus

This species is known as African horned cucumber because it is from Africa and its foliage resembles that of cucumber. However, it is not at all closely related to cucumber, but is closer in phylogeny to melon and other subgenus *Melo* species.

The oblong fruits are 10–15 cm long and have broad-based, stout spines measuring ca 1 cm in length. After about a month on the vine, the green, mottled fruits reach maximum weight, then take another 2 weeks to reach maturity, turning sweeter and orange in the process. They are quite colourful, with their bright orange skin, deep green flesh and bizarre spines. They have been said to have a flavour combining that of bananas, lemons and passion fruit, but others consider this hyperbolic and to them the fruit has an insipid or astringent taste. Fruits from indigenous wild plants are bitter. The leaves are sometimes cooked and eaten in Africa.

In recent years, New Zealand growers, encouraged by their success in marketing kiwi fruit, have exported African horned cucumber under the trade name 'kiwano' (Fig. 4.2). The fruits have a long storage life and ship well. The commercial potential of this crop continues to increase as researchers study ways to boost seed germination, stabilize fruit characteristics such as size and uniform coloration, enhance flavour, and extend storage life under humid conditions (Benzioni *et al.*, 1993).

Similar to other cultivated species of *Cucumis*, the African horned cucumber can become naturalized in areas where it is not native. For example, it has been a pest in sugarcane fields and on farms in Queensland, Australia for over 60 years.

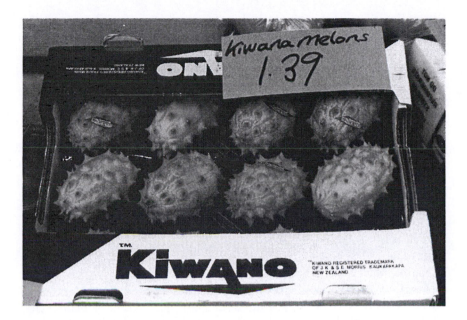

Fig. 4.2. African horned cucumber fruits from New Zealand for sale in Ireland.

Cucumber

Botany

Most cucumber plants are indeterminate, producing a small trailing vine 1–3 m long, but determinate cultivars with a compact plant habit have been developed for home growers and for mechanical, once-over harvesting. A single, unbranched tendril develops at each leaf axil. Trichomes occur on the angular stems and the triangularly ovate, 3–5-lobed leaves.

Cucumis sativus was originally monoecious, as are many modern cucumber cultivars, but gynoecious and andromonoecious cultivars were subsequently bred. Generally, only one of the clustered staminate flowers at a node is open on a given day. Male and female flowers are typically borne at different nodes, with the female flowers at higher, i.e. more distant, nodes than the male flowers. Female flowers are usually solitary at occasional nodes, but there may be several female flowers at a node if the plant has a multiple pistillate allele, and gynoecious plants have a female flower at every node. The inferior ovary has three united carpels in most cultivars, five in the 'Lemon' cultivar. Immature fruits are green at the edible stage, except in a few cultivars, where they are white or yellow. Fruits are round to oblong or narrowly cylindrical, with small tubercles (warts) and spines of trichome origin on the rind. Spine colour is associated with mature fruit colour and fruit netting. Fruits of white-spined cultivars are light green to yellow at maturity and not netted. Black-spined fruits become orange or brown when mature and may be netted. Fruit flesh is crisp and usually white, but is pale orange in a few cultivars. Seeds are small, white and flat.

Origin and history

Cucumber is of Asiatic origin; the progenitor may be the closely related, wild *Cucumis sativus* var. *hardwickii*, which was first found in the Himalayan foothills of Nepal. Plants of this variety are highly branched, daylength sensitive and prodigious producers of bitter fruits.

Cucumber remains in eastern Iran have been dated to the third millennium BC. Cucumber cultivation goes back at least 3000 years in India and 2000 years in China. China is considered a secondary centre of genetic diversification. Today, cucumber is one of the most important vegetable crops in that country, second only to Chinese cabbage in area cultivated.

Early travellers brought cucumber to Mediterranean countries 3000–4000 years ago, where the fruits were esteemed by the ancient Romans and Egyptians. In the early fourteenth century, cucumber plants were cultivated in the UK; there, the fruits were known as 'cowcumbers'. Portuguese explorers subsequently carried cucumber to West Africa. Columbus introduced this species to the New World, planting it in Haiti in 1494. Today, cucumber is grown throughout the world in small gardens, large commercial farms, and glasshouses.

Uses

There are a few non-food uses of cucumbers. White-skinned cultivars were grown in France in the nineteenth century for the production of cosmetics. Today, cucumbers are still used in some health and beauty products, including perfumes, lotions, soaps and shampoos. Indigenous practitioners create medical concoctions from the roots, leaves, stems and seeds.

The common use of cucumbers is as food. They are most often consumed fresh or pickled. In China, India, Indonesia, Malaysia and some other countries, they may be cooked before eaten. The fruits are used in curries and chutney in India. Cucumber seeds are eaten, particularly in Asia, and they yield an edible oil which is sometimes used in French cuisine. Young leaves and stems are cooked in southeastern Asia.

Cucumber cultivars are classified as slicers, usually served fresh in salads, or picklers, which are often fermented. However, in some areas, fruits of pickling cultivars are used like slicers in salads. Small-fruited pickling cucumbers are called gherkins in various countries, including India.

Generally, picklers have shorter fruits with more prominent warts than slicers. The length to width ratio, usually about three to one, is important for pickling cucumbers. Most slicing cultivars have white-spined fruits, but picklers may have either white or black spines. Pickling cultivars with white spines are becoming more popular because their fruits retain green colour longer.

After harvest in commercial operations, slicers are usually graded, washed and waxed before being marketed. Picklers are prepared for fresh pack (unpasteurized, refrigerated dills) and for brined and fermented products (see Chapter 6 and Fig. 6.10 for more details).

Over 70% of the USA cucumber crop is pickled. Yield of pickling cucumbers has increased more than three-fold in the USA in the past 60 years, from 3.61 metric tonnes per hectare in 1930 to 11.61 in 1990.

Cultivars and breeding

Glasshouse cucumber cultivation is important in northern Europe, Asia, the Middle East and other areas. Glasshouse production of cucumber in the USA has declined in recent years, however, due to competition in winter from plants grown outdoors in Florida, Mexico and other warm regions. Slicing type cucumber cultivars are most commonly grown in glasshouses. Pickling cultivars have been grown in glasshouses in The Netherlands, but this practice is declining. The cultural practices associated with glasshouse cucumbers are described in Chapter 5.

Glasshouse cucumbers have become popular in markets, commanding a premium price because of their excellent quality. They can often be identified by their very long, slender fruit, with constricted neck, thin skin, indistinct warts and spines, and crisp texture. Their crispness, tender rind and other quality attributes are primarily due to the cultivar, rather than to being grown in a glasshouse. Glasshouse cultivars include the European gynoecious 'Telegraph' and its selections, the high yielders 'Petita F_1' and 'Superator', and the small-fruited 'Hayat', which is popular in the Middle East.

Some of today's cultivars are known to be centuries old, having originally been developed in Europe or Asia. An example is 'Early Russian', which was described by Naudin in France in 1859. Many field-grown cultivars in Europe and the Middle East differ from American cultivars by having numerous fine spines and indistinct warts.

Cultivar selection in the USA began in the late 1880s, with emphasis placed on fruit shape and colour as well as adaptation to local growing conditions. Many cultivars introduced before 1900 were developed simply by selecting superior plants from the heterogeneous cultivars grown at that time. American and English cultivars were crossed to develop 'Tailby's Hybrid', introduced in 1872, and controlled hybridization has subsequently become very important in cucumber improvement.

Older American cultivars of slicing cucumbers develop stippling at high temperature, i.e. the fruit develops light green spots. Cultivars from Asia and most European glasshouse cucumbers have an allele for uniform green fruit colour, and H.M. Munger has backcrossed this allele into 'Marketmore 70' and other breeding material to produce cucumbers that retain their uniform, dark green fruit colour even at high temperatures. Most slicing cultivars marketed in the USA today have this characteristic.

'Maine No. 2', released in 1939, was the first cucumber cultivar resistant to the scab disease. Its resistance is due to a single dominant allele. Scab is no longer an important afflication of most cucumber crops because many modern cultivars possess this allele.

Cultivars resistant to CMV have been available for more than 50 years. J.C. Walker combined scab and CMV resistance in 'Wisconsin SMR 18', an important pickling cultivar introduced in 1958. A higher level of resistance to CMV, combined with good horticultural type, was achieved in the 'Marketmore' series of slicing cultivars bred by H.M. Munger. He incorporated resistance to scab and additional diseases into this cultivar by backcrossing. Cucumber breeders continue to combine genes for resistance to different diseases. The 'Sumter' cultivar is resistant to seven diseases and 'WI 2757' to nine.

'Burpee Hybrid', the first F_1 hybrid cucumber cultivar, was bred by O. Shifriss and introduced by the Burpee Seed Company in 1945. The development of hybrid cultivars became very important after gynoecious sex expression was obtained from a Korean cultivar. The gynoecious allele is dominant, and gynoecious hybrid cultivars often bear a high proportion of female flowers, resulting in earliness and good yield. They also may have good concentration of maturity, yielding many fruits in a single, mechanical harvest. 'Spartan Dawn', introduced in 1962, was the first gynoecious hybrid cultivar. C.E. Peterson developed the maternal parent of this hybrid by backcrossing the gynoecious allele into 'Wisconsin SMR 18'. The most popular pickling cultivars today are gynoecious hybrids.

'Lemon' is a unique cultivar. It has andromonoecious sex expression, whereas other cucumber cultivars are monoecious or gynoecious. The small, yellow fruit, which has five placentae instead of the customary three, faintly resembles a citrus fruit, hence its name.

A distinct group of cultivars, known as Xishuangbanna gourd, is classified as *C. sativus* var. *xishuangbannesis* Qi & Yuan. Grown by the Hani people of southwestern China at elevations of 1000 m or higher, these landraces are largely unknown outside of eastern Asia. Vines of this variety are vigorous, reaching up to 7 m in length. The large mature fruit is oblong and weighs about 3 kg. The rind is white, yellow or brownish-orange, sometimes with distinct netting, but otherwise relatively smooth and without prominent warts. Because the yellow to orange flesh colour is caused by provitamin A carotenes, researchers (e.g. Navazio, 1994) have begun experiments to transfer these nutrients from the Xishuangbanna gourd to commercial cultivars.

Large, oblong fruits are also found among cultivars of var. *sikkimensis* Hook. f. (Sikkim cucumber). These cultivars are grown primarily in mountainous Nepal and India. Both the Sikkim cucumber and Xishuang-banna gourd are quite variable and include cultivars with five-carpellate fruits.

Over the past 30 years, Chinese cucumber breeders have produced approximately 40 inbred and F_1 hybrid cultivars (e.g. 'Ningqin', 'JingYan No. 2'), which occupy 40% of the cucumber production area in China (Cui and Zhang, 1991). Selection practices have focused on gynoecy, earliness and multiple disease resistance. Breeders are also trying to improve various traits for autumn crop production. Chinese consumers demand long fruits with a dark green rind and thick, prominent spines.

Breeders worldwide continue to select for a wide range of desirable characteristics, in addition to disease and pest resistance. Cultivars have been bred for tolerance to cold, heat, drought, herbicides, sulphur dioxide and soil salinity. Pickling cucumbers have been bred to withstand carpel separation in order to prevent bloating during the brining process. Cucumber cultivars differ in the time required for their fruits to develop from the optimal size for market to oversized fruits of little value, and 'Marketer' has become important because it produces a large proportion of fruits of marketable stage (Munger, 1992). Newly developed multiple branching cultivars (e.g. 'littleleaf' types) produce more fruits at the same stage of maturity, which is desirable for once-over harvesting systems. Recent breeding efforts towards increasing yield and multiple branching have focused on the wild var. *hardwickii* (Lower and Nienhuis, 1990).

Melon

Nomenclature

Many common names have been used for this species, including muskmelon, cantaloupe, nutmeg melon, winter melon (also a name for *Benincasa hispida*), sweet melon, rock melon, snap melon and others. Munger and Robinson (1991) recommend that it simply be called melon.

Botany

Melon stems are nearly round as opposed to angular in cucumber, and the pubescence of melon is not as harsh. Plants typically are trailing vines, but compact cultivars with short internodes have been bred. Tendrils are unbranched and borne singly at the nodes. Leaves are usually subcordate and less lobed than those of cucumber. Most melon cultivars are andromonoe-cious and have round or oval fruits, but some (e.g. 'Banana') are monoecious and produce elongated fruits. Andromonoecious cultivars generally have staminate flowers in axillary clusters on the main stem and laterals, and a single perfect flower at the first node of each lateral branch. One or two fruits are usually produced per plant, but cultivars with small fruits may have more. Immature fruits are pubescent; mature fruits are glabrous. The small, white to tan seeds are oblong to elliptical.

The fruits of different cultivars are quite diverse, making melon the most variable species in its genus. Some cultivars have fruits with a reticulate netting on the rind, whereas others have smooth rinds. Indented vein tracts distinguish various cultivars, but these sutures are lacking in other cultivars. Rind colour varies from green, sometimes with white stripes, to yellow, tan or white when mature. Flesh colour may be orange, pink, green or white. An abscission layer forms at the attachment of the fruit to the peduncle in some cultivars, causing the fruit to separate from the vine at maturity. In other cultivars, the peduncle is persistent.

Melon plants will not cross with cucumber plants, although these species are members of the same genus. The mistaken belief that cucumber and melon are able to cross may have arisen from harvesting melons when immature or after their quality has been impaired by disease or unfavourable climate, and assuming that the insipid flavour of the melon was due to cross-pollination with a cucumber. Cucumber flowers can be pollinated by melon pollen, but the embryo aborts before it can be rescued by embryo culture techniques. Seedless cucumbers may be produced from cross-pollination with melon pollen, but melon fruits generally do not develop when pollinated with cucumber pollen.

Origin

The origin of melon is in dispute. Some suggest that India is the centre of domestication because melon has been cultivated there for centuries and inedible forms grow wild in India. Others believe that melon domestication began in Iran. Most authorities, however, consider that melon, like other species in the subgenus *Melo*, originated in Africa.

Historical records and archaeological remains place melon in Egypt and Iran during the second and third millennia BC, respectively. Melon spread throughout the Middle East and Asia, becoming an important vegetable in India, Egypt, Iran and China. India, Iran, Afghanistan and China are considered secondary centres of melon diversification. Genetic diversity in Spain is great as well.

In the fifteenth century, melon was brought from Turkish Armenia to the papal estate of Cantaluppe near Rome, and was distributed from there to western Europe. The word 'cantaloupe' is derived from this source; however, the small melon with a hard rind grown in Italy at that time is quite unlike the modern cultivars known as cantaloupes today. Ancient Romans who grew melon did not prize the fruits as much as they did cucumbers, presumably due to the poor quality of the melon cultivars they grew. In the seventeenth century, melon cultivars were under cultivation in British glasshouses.

Melon was brought to the New World by Columbus and carried to California by the Spaniards in 1683. Most of the early American cultivars had green flesh. 'Rocky Ford', a selection of the green-fleshed 'Netted Gem', became important in the 1800s. Orange-fleshed cultivars such as 'Bender' and 'Irondequoit' were popular near the turn of the century. Improved orange-fleshed cultivars, including 'Hearts of Gold' (introduced in 1914) and 'Hales Best' (1924), were predominant for many years, until they were replaced by disease resistant cultivars in the 1930s. 'Burpee Hybrid', the first F_1 hybrid melon, was introduced in 1955, and hybrid cultivars have subsequently become important.

Melon cultivars are now grown in temperate and tropical areas throughout the world. China produces more than twice as many fruits as any other country (Table 2.1). Melons are so prized in Japan that they are grown in glasshouses, command prices many times those prevailing in other countries, and are often given as gifts. Western Asia (e.g. Turkey and Iran) and northern Africa (e.g. Egypt and Morocco) are also important areas for melon production. Spain, Romania, Italy and France are the largest European producers (FAO, 1995).

Classification, cultivars and uses

French botanist Charles Naudin determined in the mid-nineteenth century that several *Cucumis* species would cross readily with each other and with melon, and he combined them as different botanical varieties of *Cucumis melo*. Today, with slight modification, Naudin's categories are considered horticultural groupings based on fruit characteristics and uses, and not botanical varieties based on phylogeny (Munger and Robinson, 1991), nor conforming necessarily with taxonomic rules for nomenclature (Trehane *et al.*, 1995). These cultivar-groups are:

1. Cantalupensis Group (common names are cantaloupe and muskmelon). Medium-sized fruits with netted, warty or scaly surface; flesh usually orange but sometimes green; flavour aromatic or musky. Fruits detach from peduncle at maturity. Usually andromonoecious.
2. Inodorus Group (winter melon). Fruits usually larger, later in maturity and longer keeping than those of the Cantalupensis Group. Rind surface smooth or wrinkled, but not netted; flesh typically white or green and lacking a musky odour. Fruits do not detach from the peduncle when mature. Usually andromonoecious.
3. Flexuosus Group (snake melon). Fruits are very long, slender and often

ribbed. They are used when immature as an alternative to cucumber. Monoecious.

4. Conomon Group (pickling melon). Small fruits with smooth, tender skin, white flesh, early maturity and usually with little sweetness or odour. They are often pickled and are also eaten fresh or cooked. Andromonoecious.

5. Dudaim Group (pomegranate melon, Queen Anne's pocket melon), which includes members of the previously recognized Chito Group (mango melon, vine peach). Small, round to oval fruits with white flesh and thin rind.

6. Momordica Group (phoot, snap melon). Fruits are ovoid to cylindrical in shape, with dimensions of 30–60 × 7–15 cm. Flesh is white or pale orange, low in sugar content, mealy, and insipid or rather sour tasting. The smooth surface of the fruit cracks or bursts as maturity approaches and the fruit disintegrates when barely ripe. Most of the cultivars are monoecious.

These cultivar-groups belong to ssp. *melo*. Although plants of ssp. *agrestis* (Naud.) Pang. are mostly wild, slender vines with small, inedible fruits, they are sometimes cultivated in Asia and occur as weeds in many tropical areas.

Cantalupensis Group cultivars are the most important in commerce. Cultivars grown in the western USA that are called cantaloupes produce moderately ribbed, tightly netted fruits which ship well and have thick, orange flesh. 'PMR 45' was introduced by the USDA and the University of California in 1936, and many selections of this cultivar were subsequently made by seedsmen. Because of their resistance to powdery mildew, 'PMR 45' and its derivatives have dominated California melon production for more than half a century. 'Persian' and related cultivars, which are also grown in California, are similar to 'PMR 45', but have larger fruits with heavy netting. Cultivars grown in the eastern USA may also produce large fruits, but generally with less netting and more ribbing (e.g. 'Iroquois'). High-quality melons of the Charentais type (e.g. 'Charentais', 'Charmel', 'Alienor'), with thin rinds and only sparse netting, are popular in Europe and Africa. Green-fleshed muskmelons developed in Israel (e.g. 'Ogen') are also popular in Europe.

Breeders have surveyed and selected for various traits in Cantalupensis Group cultivars. 'Top Mark', a leading cultivar in the USA today, is tolerant of environmental conditions detrimental to many other cultivars: it can be grown on saline soil; it tolerates high concentrations of atmospheric ozone; and it can be treated with sulphur at high temperature for disease control without injury. Melon cultivars have also been found to differ in sensitivity to high boron concentrations in soil. Birdsnest-type cultivars from Iran, which have a compact plant habit, reduced apical dominance, good fruit set and concentration of maturity, are potentially valuable for once-over harvesting (McCollum *et al.*, 1987). 'Siberian Honey Dew', an Inodorus Group cultivar, has been used as a parent to breed long storage life of fruits into cultivars of the Cantalupensis Group.

Cultivars of the Inodorus Group are called winter melons because the fruits generally store well. Some are shipped thousands of miles to distant markets from Spain, California and other areas with a long, warm growing season. 'Honey Dew', the most popular member of this cultivar-group, was introduced to the USA from France at the beginnning of the twentieth century. The firm fruit has green, very sweet flesh. It is generally grown in the western states of the USA and other long-season areas because of its late maturity. Bush habit has been backcrossed into 'Honey Dew' and related cultivars. Another winter melon cultivar, 'Casaba', has a yellow, furrowed rind and white flesh of high quality. It also requires a long growing season. 'Juan Canary' fruits are yellow and elongated, with light green flesh. 'Crenshaw' has delicious, juicy, orange flesh, but does not ship or store as well as 'Honey Dew', 'Casaba' or 'Juan Canary'. 'Honey Gold' is one of the more recently created orange-fleshed winter melon cultivars. Various Spanish landraces (e.g. 'Común') of the Inodorus Group have exhibited resistance to powdery mildew (Floris and Alvarez, 1995).

Some Flexuosus Group melons are very long (up to 150 cm in 'Cooking Queen'), slender and coiled (less so when plants are trellised), hence their common name of snake melon (Fig. 4.3). The 'Armenian Cucumber' cultivar listed in seed catalogues is not a cucumber, but a melon of this cultivar-group. Snake melons are popular in the Middle East, where they are harvested when immature and consumed like cucumbers. They are better than cucumber plants at setting fruit at the high temperatures often prevailing in these countries.

Pickling melon cultivars of the Conomon Group produce small, crisp fruits with white flesh and often a low sugar content. This lineage is believed to have originated in China and Japan more than 1000 years ago. The fruits are typically pickled there, and are said to make better quality pickles than cucumbers. They have a very thin, tender rind, which turns white when mature in a few cultivars (e.g. 'Shiro-Uri'). Cultivars such as 'Golden Crispy' may be eaten fresh with skin intact, like an apple. Young fruits are also cooked, in ways similar to summer squash. Resistance to CMV has been found in these cultivars, and breeders are attempting to transfer this resistance to members of the Cantalupensis Group.

Some Dudaim Group cultivars are grown as ornamentals because of their aromatic fruits, but others lack this fragrance. Fruits are eaten fresh, though they are not very sweet, or pickled. The flesh resembles that of a cucumber in colour and texture. Plants of this Asian cultivar-group have escaped from cultivation and become established as weeds in southern USA and elsewhere.

Momordica Group cultivars are grown in India and other Asian countries, where the fruits are cooked as a vegetable, consumed fresh or used in salads. The seeds are roasted and eaten in India. A USDA plant introduction (PI) from that country, which was sold by a street vendor for roasting, included both cucumber and melon seeds. The melon germplasm of that introduction, designated PI 175111, has been very useful in melon breeding

Fig. 4.3. Fruit of a Flexuosus Group melon measuring approximately 1 m long.

programmes because of its resistance to aphids and viruses.

Melon seeds, often dried and ground, are commonly eaten in Africa, where they can be found in commercial trade among countries. Oil from the seeds is used for illumination. Leaves are sometimes prepared as a vegetable, and the whole plant provides fodder.

In addition to food, various parts of the melon plant have been used medicinally. In China, fruits and roots are taken as a diuretic, roots and flowers as an emetic, leaves and seeds for the treatment of hematoma, and stems for the treatment of dysentery and hypertension. Melon plants contain various bioactive principles, including elaterin, stigmasterol, spinosterol and the antitumor cucurbitacin B (Duke and Ayensu, 1985).

SQUASH

Botany

In *Cucurbita*, there are five domesticated (*C. argyrosperma*, *C. ficifolia*, *C. maxima*, *C. moschata*, *C. pepo*) and approximately ten wild species (Tom Andres, New York, 1996, personal communication). Most species, including the domesticated species, are mesophytes with fibrous root systems; the remaining taxa are xerophytic perennials with enlarged roots. All produce frost-sensitive, tendril-bearing vines. Wild species and most cultivars have long, trailing vines, but some cultivars, particularly in *C. pepo*, have a compact bush habit in which the tendrils have been reduced in size and function. The large leaves are palmately lobed to nearly round. Some cultivars of *C. pepo* and *C. maxima*, as well as many cultivars of *C. moschata* and *C. argyrosperma*, have mottled leaves, with white or silvery areas at the junctions of principle veins.

Species of *Cucurbita* are monoecious. The unisexual flowers are large, showy and orange (except for the cream-flowered *C. okeechobeensis*), occurring singly in leaf axils. The five petals are reflexed at their tips and fused at their bases. Calyx lobes are narrow (*C. pepo*) or broad to sometimes leaf-like (*C. moschata*). Petals alternate with sepals, which are also united at their bases and fused with the lower corolla to form a cup-like hypanthium. Although the three filaments are separate, the anthers are more or less united and produce abundant amounts of heavy, sticky pollen. Styles are typically joined together, but they diverge slightly where the stigmas are attached. Nectar is produced in a disc inside and at the base of the hypanthium (Fig. 1.3). The unilocular, inferior ovary has three or five placentae, corresponding to the number of bilobed stigmas. The fruit is a pepo, and there is a great diversity in fruit size, shape and colour among cultivars. Fruit flesh is bitter in wild taxa and in most ornamental gourds of *C. pepo*. Seeds of the domesticated species are large, measuring up to 3 cm long in *C. argyrosperma*.

Fruits of the wild species have hard, lignified rinds which help to protect the seeds from herbivores. The fruit may remain intact long after the plant has died. After lengthy storage, little is left except for the dried rind, peduncle, and seeds; these fruit parts may persist for centuries, enabling archaeologists and botanists to determine prehistoric species distributions and uses. The intact dried fruits are buoyant, permitting seed dispersal via waterways.

Nomenclature

Different *Cucurbita* species are known by the same common name, and different common names, such as squash, pumpkin, cushaw and gourd, have been used for the same species. We will use the unqualified term squash to refer to any or all of the domesticated species of *Cucurbita*.

The word squash is derived from the American aboriginal word 'asku-tasquash', meaning eaten raw or uncooked. Cultivars are classified as summer squash (sometimes called vegetable marrow) or winter squash, depending on whether the fruit is used when immature or mature. The term winter squash refers to the ability of the fruit to be stored until the winter months. Summer squashes are generally *C. pepo*, but winter squashes may be *C. pepo* (e.g. 'Acorn'), *C. maxima* ('Hubbard'), *C. moschata* ('Butternut') or *C. argyrosperma* ('Green Striped Cushaw'). In *C. pepo*, bush habit distinguishes most summer squash cultivars from winter squashes.

Pumpkin comes from the old English word 'pompion', the Greek 'pepon' and the Latin 'pepo', which together mean a large, ripe, round melon or gourd. Today, the term pumpkin is used in various ways and has no botanical meaning. It typically refers to any squash used for pies, jack-o'-lanterns or stock feed. *Cucurbita maxima* and *C. moschata* cultivars that would be called winter squashes in the USA are often called pumpkins in India and other countries.

Cushaw defines a winter squash cultivar with a curved neck. Its use is not limited to a single species. Thus, 'Green Striped Cushaw' is *C. argyrosperma*, but 'Golden Cushaw' is *C. moschata*.

Gourd often designates a cucurbit not used for food, e.g. wild species of *Cucurbita*. The cultivated ornamental gourds of *C. pepo*, which have small fruits of a wide assortment of shapes and colours, are used for decoration. *Cucurbita* gourds have hard shells, and some consider this a distinction between gourd and squash. However, the fruit of some summer squash cultivars also has a hard rind when mature.

Origin and History

On the basis of isozyme banding patterns, Puchalski and Robinson (1990) classified *Cucurbita* species into seven phylogenetic groups. *Cucurbita moschata* and *C. argyrosperma* were included in the same group, and the other cultivated species were placed into different groups. In spite of their occurrence in the same phylogenetic lineage, *C. moschata* and *C. argyrosperma* were probably domesticated from distinct wild progenitors, as were the other cultivated squashes.

Most *Cucurbita* species originated in Mexico, but a few species, including *C. maxima*, are native to South America. Since the wild fruits of this genus are bitter and inedible, early gatherers probably first collected the fruits for their edible seeds or to use the durable rinds as containers. These uses eventually gave way to the domestication of species with edible fruits. From these beginnings, squash, along with corn and beans, became a staple in the diets of the Aztec, Incan and Mayan civilizations of Latin America.

Archaeological evidence places wild populations of *C. pepo* in Mexico and eastern USA around 10,000 and 30,000 years ago, respectively. Domestication of this species apparently took place independently in these two areas (Decker, 1988). The USA cultivars of ssp. *ovifera* (L.) Decker var. *ovifera*, which include various summer squashes and most ornamental gourds, were probably selected mainly from wild populations of var. *ozarkana* Decker-Walters inhabiting the Mississippi Valley. However, populations in Texas (var. *texana* (Scheele) Decker) and northeastern Mexico (ssp. *fraterna* (Bailey) Andres) may have contributed to the genetic evolution of these cultivars as well. The wild progenitor of the Mexican lineage of cultivars (ssp. *pepo*), which includes pumpkins and vegetable marrows, is currently unknown and possibly extinct (Decker-Walters *et al.*, 1993).

Cucurbita pepo was the first squash introduced to Europe. Some of the fruits portrayed in ancient European herbals are not unlike those of modern cultivars. Secondary diversification of pumpkin and vegetable marrow cultivars occurred in Asia Minor. Today, *C. pepo* is grown throughout the world.

Cucurbita moschata was cultivated in Mexico, South America and the southwestern USA in pre-Columbian times. It may have been domesticated independently in Mexico and northern South America; unfortunately, the wild ancestor(s) of *C. moschata* are currently unknown. As a cultivated plant, this species migrated throughout the Caribbean islands, giving rise to various indigenous 'calabaza' landraces. When it reached Florida, Native Americans developed a distinct landrace called 'Seminole Pumpkin'. Additional diversification of cultivars has taken place in several areas of Asia and Africa.

Cucurbita argyrosperma was apparently domesticated in southern Mexico, where archaeological evidence of cultivation dates back to 5200 BC. Wild populations of ssp. *sororia* (Bailey) Merrick & Bates, which occur in Mexico and Central America, gave rise to the domesticated subspecies (ssp. *argyrosperma*). Landraces of ssp. *argyrosperma* var. *argyrosperma* may have been selected first; the other two groups of cultivars, var. *callicarpa* Merrick & Bates and var. *stenosperma* (Pang.) Merrick & Bates, probably evolved from northern and southern landraces of var. *argyrosperma*, respectively (Merrick, 1990). Weedy populations of var. *palmeri* (Bailey) Merrick & Bates are considered escapes of var. *callicarpa* that may have undergone introgression with ssp. *sororia*. Today, *C. argyrosperma* is cultivated in Mexico, Latin America, the Caribbean, southwestern USA, and Asia, but is less commercially important than *C. pepo*, *C. maxima* or *C. moschata*.

Cucurbita maxima ssp. *andreana* (Naud.) Filov, a bitter-fruited native of Argentina and Uruguay, is the wild progenitor of domesticated ssp. *maxima*, and the two taxa hybridize readily. This species did not reach North America until it was brought by sailing ships from South America to northeastern USA in the eighteenth century. Since then, it has been especially popular in New England and is consumed in other parts of the USA as well. Today, *C. maxima* is cultivated worldwide, particularly in South America, India and Africa. Because this species is generally more cold tolerant than *C. argyrosperma*,

C. moschata and *C. pepo*, it can be grown at higher latitudes and elevations.

Cucurbita ficifolia is called the Malabar gourd because it once was thought to be of Asiatic origin. However, like other squashes, its origin is in the Americas, probably in the high altitudes of Mexico. This cold-tolerant species, growing at elevations of up to 2600 m, was also cultivated in the Andes Mountains of Peru in prehistoric times. It reached Asia on sailing ships shortly after the Discovery; the long-keeping fruits, which can be stored for up to two years, could withstand the long voyages. Malabar gourd is seldom grown in industrialized countries, except as a rootstock on which to graft cucumber. In Latin America, however, this species provides a vegetable called 'zambo' as well as food for livestock; even the fruits of wild plants are gathered and eaten (Andres, 1990).

Species Descriptions

Squash species can be distinguished on the basis of peduncle, foliage and seed characteristics (Table 4.1). The peduncle is particularly distinctive for each of the domesticates.

Cucurbita pepo is a highly polymorphic species, including various summer squashes (vegetable marrows, crooknecks, straightnecks, zucchinis, coco-zelles and scallops), ornamental gourds, winter squashes (acorns), pumpkins and unique cultivars such as 'Vegetable Spaghetti' (Fig. 4.4). Its fruit types are more genetically diverse than those of any other species in the *Cucurbitaceae*. They are green, yellow, orange, white, striped or variegated; smooth, ribbed, furrowed or warty; and flat, round, oval, elongated, necked or otherwise shaped. Fruit size ranges from small (5 cm in diameter in some ornamental gourds) to large (over 50 cm in diameter in some pumpkins). Flesh colour is usually white for summer squash and orange for winter squash, but often of a lighter intensity than the orange flesh colour of winter squashes of *C. maxima* and *C. moschata*.

Cultivars of *C. pepo* ssp. *ovifera* generally have smaller reproductive and vegetative parts than those of ssp. *pepo*. Fruits of wild populations are small, bitter, round to pyriform, and typically white (ssp. *ovifera* var. *ozarkana*), green and white striped (var. *texana*), or green and white striped turning yellow-orange when mature (ssp. *fraterna*).

Some *C. maxima* cultivars produce the largest of all squashes. Fruits of this species are orange, green or grey; smooth or ribbed; typically round or oval; and sometimes with a protuberance at the blossom end. Seeds are usually quite large and plump, white or brown, and rugose or smooth.

A squash species used for food in Mexico, Latin America and southwestern USA for centuries was not recognized as a distinct species until 1930, when a Russian botanist named it *C. mixta* Pangalo. Before then, it was included in *C. moschata*, a species morphologically similar, but differing in

Table 4.1. Distinguishing characters for domesticated species of squash (*Cucurbita*).

Species	Peduncle	Stem	Leaf	Seed
C. argyrosperma	Hard, angular but becoming round at maturity, corky, only slightly flared at fruit attachment	Hard, angular, grooved	Moderately lobed, softly pubescent	Usually white, may be very large; surface smooth or split; margin prominent, smooth to ragged, and sometimes dark
C. ficifolia	Hard, smoothly angled, with slight flaring	Hard, smoothly grooved	Lobed, nearly round, slightly prickly	Usually black, sometimes tan; surface often minutely pitted; margin smooth and narrow
C. maxima	Generally soft, round, often corky, not flared	Soft, round	Usually unlobed, nearly round, soft	White to brown, often plump; surface sometimes split or wrinkled; margin very narrow; seed scar oblique
C. moschata	Hard, smoothly angled, broadly flared	Hard, smoothly grooved	Nearly round to moderately lobed, soft	Dull white to brown; surface smooth to somewhat rough; margin prominent, typically ragged, and often dark; seed scar slightly oblique
C. pepo	Hard, angular, sometimes slightly flared right next to fruit	Hard, angular, grooved, prickly	Palmately lobed, often deeply cut, prickly	Dull white to tan; surface smooth; margin prominent but usually smooth; seed scar square or rounded

isozymes and separated by sterility barriers. *Cucurbita mixta* was recently reclassified as *C. argyrosperma*, and the former species *C. palmeri* Bailey and *C. sororia* Bailey were designated as infraspecific taxa of *C. argyrosperma* (Merrick and Bates, 1989).

Variety *argyrosperma* and var. *stenosperma* have round to pyriform, cream-coloured fruits with green-mottled stripes. Fruits of var. *callicarpa* are more variable in shape and coloration. Those of var. *palmeri* and ssp. *sororia*

Fig. 4.4. Fruits of *Cucurbita pepo*. From upper left, going clockwise: 'Gourmet Globe', 'Vegetable Spaghetti', 'Pankow's Field', 'Small Sugar', 'Yellow Crookneck', 'Vegetable Gourd', 'Table Queen', 'White Bush Scallop', 'Scallopini', 'Gold Rush'. (Whitaker and Robinson, 1986. Reprinted courtesy of Chapman & Hall, New York.)

are relatively small, often with bitter flesh, especially in ssp. *sororia*. The pale yellow to orange flesh of the cultivars is coarse and often of rather poor quality for a winter squash. A dark green tint to the placental tissue is unique to ssp. *argyrosperma* var. *stenosperma*. Variety *argyrosperma* has very large seeds, which are often eaten in Mexico and Guatemala. Variety *stenosperma* is also grown primarily for its edible seeds. Seeds of var. *callicarpa* are variable in colour and surface characteristics. The tan seeds of ssp. *sororia* are small, whereas the white seeds of ssp. *argyrosperma* var. *palmeri* are somewhat larger.

    Cucurbita moschata cultivars produce fruits that are often large; smooth, warted, or wrinkled; sometimes mottled; and typically green or buff-coloured. Landraces have flattened, round, oval, or curved-neck fruits. Breeders have developed types having a long, straight neck with the small seed cavity confined to the enlarged blossom end of the fruit, e.g. 'Waltham Butternut'. Winter squash cultivars of this species store well and can be of very high quality.

    The species name for *C. ficifolia* and its common name of fig leaf gourd refer to the resemblance of its leaves to the shape of fig tree leaves. Despite its long history of cultivation, there is little genetic diversity in Malabar gourd;

fruits are white or green and white, with white, coarsely fibrous flesh (Andres, 1990). In shape, the fruit resembles a watermelon. No other wild plants of a domesticated squash species produce fruits as large as those of feral Malabar gourd. This is the only species of *Cucurbita* with black seeds, although landraces with tan seeds also occur.

Malabar gourd is a short-day plant, and it may not flower early enough to produce fruits outside of the tropics. Seedsmen producing seed of this species for use as rootstock have selected for earlier flowering, day-neutral forms, but even these selections are late to mature.

Uses

During domestication in pre-Columbian times, people selected for non-lignified fruit rinds in winter squashes and pumpkins, which were usually consumed when ripe. Hard rind, which is conditioned by a single dominant allele in *C. pepo*, is found in many cultivars of summer squash that are consumed before rind lignification becomes objectionable.

Native Americans dried strips of squash flesh in the sun for preservation. Today, summer squash is usually cooked by boiling or frying, and winter squash by baking, boiling or microwaving. Summer squash generally has white flesh and low soluble solids content, but winter squash has been bred to have orange, carotenoid-rich flesh high in soluble solids.

Immature fruits of winter squash cultivars like 'Jersey Golden Acorn' (*C. pepo*) may be eaten like summer squash. 'Butternut' and similar cultivars of *C. moschata* are of high quality when immature. However, they are seldom marketed this way due to relatively low yield, late maturity, and the viny habit, which makes harvesting difficult. In Argentina, bush cultivars of *C. maxima* known as 'zapallito' are a popular form of summer squash.

Some summer squash cultivars, e.g. the vegetable marrows (*C. pepo*), are consumed when almost mature. In the Middle East, nearly mature fruits of 'Cousa' are stuffed with meat and other ingredients, then baked.

In the USA, the most familiar use of pumpkins of *C. pepo* is for Halloween jack-o'-lanterns. Every autumn, the orange, round or oval fruits are carved into grotesque faces and illuminated from within by candles.

Commercially canned pumpkin pie mix may be made of *C. pepo*, *C. maxima* or *C. moschata*. Although *C. pepo* has a long tradition of use in the USA, fruits of the other two species produce better baked pies. Cultivars of *C. moschata* and *C. maxima* are also processed and sold as canned or frozen winter squash. Canners prefer cultivars with an orange rind; if a small piece of the skin is inadvertently included in the canned product, it is less noticeable than with green-fruited cultivars. Orange, carotene-rich fruits of *C. maxima* that are also fine in texture and flavour are mashed into jarred baby food.

Mature fruits of *C. ficifolia* are used as winter squash, and the immature fruits like summer squash. An Aztec candy called 'angel's hair' is prepared from the boiled stringy flesh. The bland fruits contain a proteolytic enzyme that may be of future commercial value to the food industry. Stem tips and leaves are used as greens, flowers serve as a condiment, and seeds are roasted and eaten.

Squash has been an important food in Mexico for centuries. Fruits are processed and consumed in a variety of ways. Squash seeds are a popular snack food, and are also ground into a meal used to make special sauces. The flowers are eaten stuffed or fried, and lend colour and flavour to soups, stews and salads. Stems, growing tips and roots of some species are consumed also.

New World aborigines have used squash seeds as a source of cooking oil and medicinally as a diuretic, antipyretic and anthelmintic. The fruit flesh of wild species generates a saponin-rich froth when rubbed, and has been employed to wash clothes.

Cultivars and Breeding

Because of the great diversity of cultivars in *C. pepo*, various horticultural classifications have been proposed over the years. However, a nomenclaturally correct classification of cultivar-groups (cf. Trehane *et al.*, 1995) has not yet been established for any squash species, nor will we attempt to define such here. For *C. pepo*, we will use for reference the classification by Paris (1986), which is based primarily on fruit shape; his groupings of cultivars are:

1. Pumpkin. Fruit orange, round or oval.
2. Scallop. Fruit small, flattened, typically with scalloped edges.
3. Acorn. Fruit small, top-shaped, furrowed, pointed at the blossom end.
4. Crookneck. Fruit elongated with a curved neck.
5. Straightneck. Fruit cylindrical with a straight, slightly constricted neck.
6. Vegetable marrow. Fruit short, cylindrical, tapering from the broad blossom end to the narrow peduncle end.
7. Cocozelle. Fruit long, cylindrical, tapering away from the large blossom end, with a length to width ratio of 3.5 or more.
8. Zucchini. Fruit long, cylindrical, with little or no taper.

The pumpkin, vegetable marrow, cocozelle, and zucchini groups belong to ssp. *pepo* and the remaining groups represent ssp. *ovifera*. Not included above are the ornamental gourds, which are small when mature and occur in a variety of shapes and colours.

Dark green zucchini-type cultivars (e.g. 'Black Zucchini', introduced in 1931) are the most popular summer squashes in the USA, but lighter green cultivars (e.g. 'Cousa') are preferred in the Middle East. Zucchini cultivars have long been important in Italy and the earliest selections probably originated there. F_1 hybrids, such as 'Black Jack', are popular today. Although most zucchinis have green rinds, scientists have recently selected types with the *B* allele for yellow rind (e.g. 'Burpee's Golden Zucchini').

'Table Queen', one of the few winter squash cultivars of *C. pepo*, was introduced in 1913 by the Iowa Seed Company. This acorn squash produces small, ribbed fruits which are often cut in half and baked. It is similar to a landrace grown by Native Americans before the Discovery. 'Table Queen' and its antecedents have a vining plant habit, but 'Table Ace' and other bush cultivars with fruit shape similar to 'Table Queen' have been bred.

'Delicata', another winter squash, was introduced by the Peter Henderson Company in 1894 and is still grown today. The cylindrical fruits, which are cream-coloured with green stripes, are of good culinary quality.

'Cocozelle' was introduced by the Asgrow Seed Company in 1934, and 'Caserta' by the Connecticut Agricultural Experiment Station in 1949. They are popular summer squashes with striped fruits. 'Caserta' produces a high proportion of female flowers early in the season, one reason for its use as the maternal parent of F_1 hybrid cultivars.

'Yellow Crookneck' and similar cultivars with a curved neck and hard, warty rind when mature are popular in southern USA. Cultivars of the straightneck group were probably originally selected from these crookneck squashes. 'Early Prolific Straightneck', an important yellow-coloured summer squash, was introduced by the Ferry Morse Seed Company in 1938. 'Yankee Hybrid', a straightneck squash introduced in 1942, was the first F_1 hybrid squash cultivar.

The fruit of 'White Bush Scallop' summer squash is similar to that depicted in ancient European herbals. It differs from the medieval squash, however, by having a bush habit.

'Connecticut Field' pumpkin has been grown in the USA since Colonial times and is still listed in some seed catalogues today. Miniature pumpkins of *C. pepo* (e.g. 'Jack Be Little', 'Munchkin' and 'Baby Bear'), with fruits looking like 'Jack O'Lantern' but much smaller, have become popular in recent years. They are used for decoration and are edible as well.

Vegetable marrows, which are grown in the UK and elsewhere, are used at all stages of maturity. 'Vegetable Spaghetti', an unusual member of this group, is said to have originated in Manchuria and was introduced into North American commerce in 1936. When mature, the yellow, oval fruit can be cooked intact, after which, it is cut open to serve the thin strands of flesh resembling spaghetti. Scientists from Israel selected an orange hybrid cultivar called 'Orangetti' in 1986. In addition to having a higher concentration of carotenoids in the orange flesh, this cultivar differs from the viny 'Vegetable Spaghetti' by having a semi-bush habit.

A single allele plus modifier genes in *C. pepo* inhibit formation of the seed coat. Pumpkin cultivars such as 'Lady Godiva' and 'Triple Treat' have been bred to have 'naked seeds' that are tasty, tender, and nutritious, being high in protein and oil content. Since the seed coat is maternal tissue, cross-pollination of a naked seeded cultivar will not affect that trait until the next generation.

Three horticultural groupings of *C. moschata* cultivars are recognized in the commercial trade of North America:

1. Cheese. Fruit variable, but usually oblate with a buff-coloured rind.
2. Crookneck. Fruit round at the blossom end with a long straight or curved neck.
3. Bell. Fruit bell-shaped to almost cylindrical.

These groups do not encompass all of the fruit types that have evolved in tropical America (e.g. calabaza landraces) and Asia. For example, Colombian landraces have small fruits with dark seeds, and Japanese cultivars (e.g. 'Chirimen', 'Kikuza') often have warty and wrinkled fruits.

The bell squash 'Butternut' is an important winter squash with excellent quality. It was selected for better fruit shape from the heirloom cultivar 'Canada Crookneck' and introduced by the Breck Seed Company in 1936. The elongated neck of the buff-coloured 'Butternut' fruit is generally straight but occasionally curved. The neck is entirely usable because the small seed cavity is confined to the bulbous base of the fruit. 'Waltham Butternut' is similar, but produces a higher proportion of fruits with straight necks. It was obtained by crossing 'New Hampshire Butternut' with an African plant introduction to the USA, and has been a very popular cultivar ever since its commercial introduction in 1970.

'Cheese', one of the oldest *C. moschata* cultivars grown in the USA, has a flattened, ribbed fruit similar in shape to a cheese box, with buff-coloured rind and deep orange flesh. It was popular for canning and stock feed during the nineteenth century.

After *C. pepo*, *C. maxima* exhibits the greatest diversity of fruit types (Fig. 4.5). As with the horticultural classification of *C. moschata*, not all of the localized landraces of *C. maxima* that have evolved can be placed in the following informal classification scheme, which is based on Castetter (1925):

1. Banana. Fruit long, pointed at both ends, with a soft rind and brown seeds.
2. Delicious. Fruit turbinate, shallowly ribbed, with a hard rind and white seeds.
3. Hubbard. Fruit oval, tapering to curved necks at both ends, with a very hard rind and white seeds.
4. Marrow. Fruit oval to pyriform, tapering quickly at the apex and gradually towards the base, with white seeds.
5. Show. Fruit large, orange, with a soft rind and white seeds.
6. Turban. Fruit turban-shaped as a result of fruit tissue at the blossom end not covered with receptacle tissue.

Some cultivars, such as the heavily warted 'Victor', were produced by hybridizing cultivars of different groups. The parentage of 'Victor' is believed

Fig. 4.5. Fruits of *Cucurbita maxima*. From upper left, going clockwise: line NK580, 'Mammoth King', 'Golden Delicious', USDA PI 458741, 'Turk's Turban', 'Buttercup', 'Queensland Blue'. (Whitaker and Robinson, 1986. Reprinted courtesy of Chapman & Hall, New York.)

to include hubbard and turban squashes (Tapley *et al.*, 1937).

'Buttercup' is a high-quality winter squash. Its small, dark green fruit has a 'button', a protuberance at the blossom end where the mature ovary is not covered by the receptacle. This turban group cultivar was bred by A.F. Yeager and released in 1931. 'Buttercup' plants, like those of most cultivars of *C. maxima*, are large vines. Yeager later crossed 'Buttercup' with a bush-habit USDA plant introduction of *C. maxima* and selected for bush plants with 'Buttercup'-type fruit set close to the crown. The best of these selections was named 'Bush Buttercup'.

'Turk's Turban' is a turban squash with very colourful fruits, which are used for decoration. It is consumed as a winter squash.

'Hubbard' produces a large, oval fruit that stores well and is of good quality. It was introduced by James Gregory of Marblehead, Massachusetts in 1856, but was probably originally brought to New England from South America in the eighteenth century. Cultivar selections have been made that produce orange, green or grey fruits.

'Queensland Blue' is a high-quality Australian cultivar. The flattened, deeply ribbed fruits have a bluish-grey rind.

Show pumpkins are grown for forage in India, where there is much

Fig. 4.6. Commercial seed packet of 'Early Yellow Summer Crookneck' (*Cucurbita pepo*) from the 1920s.

diversity. In various countries, cultivars producing massive, orange fruits, such as 'Atlantic Giant' and 'Big Max', are cultivated for entry into contests for large fruits.

Many squash cultivars were described and illustrated by Tapley *et al.* (1937). Some of these older cultivars still exist today (Fig. 4.6). 'Hubbard', for example, has been grown for over 150 years and 'Butternut' has been popular for more than 50 years. The situation is less stable for *C. pepo*, where there has been a rapid turnover in popularity of F_1 hybrid summer squash cultivars. However, some older *C. pepo* cultivars, such as 'Acorn', 'Scallop' and 'Connecticut Field', all more than 100 years old, are still grown today. Descriptions of many of the more modern hybrids and cultivar selections can be found in Facciola (1990).

A major and relatively recent focus of many squash breeding programmes is disease resistance. Relative to the other major cucurbit crops, squash is very vulnerable to attack by many viruses, fungi and bacteria. In addition to traditional plant breeding methods, including interspecific hybridizations to transfer disease resistance alleles from wild to cultivated species of *Cucurbita*, scientists are genetically engineering viral recognition genes and transferring them via a transgenic vector to squash, where they confer resistance (for more details, see Chapter 3). 'Freedom II' is a recently created crookneck cultivar (*C. pepo*) engineered with resistance to WMV and ZYMV.

Breeders continue to improve flesh quality in winter squashes, including increasing orange coloration and provitamin A carotene content. A 1996 introduction with very high carotene content is the F_1 hybrid 'Jade A' (*C. maxima*). Selection is also made for flavour, texture, consistency and freedom from bitterness.

Oved Shifriss transferred allele *B* from a *C. pepo* ornamental gourd to a summer squash. In the gourd genetic background, the allele produces bicoloured fruit, each fruit on a plant having a variable proportion of green and yellow rind colour. By selection for *B* and the proper modifiers, it has been possible to breed summer squash cultivars with entirely yellow fruits that are relatively high in vitamin A content for a summer squash. Symptoms of WMV may be masked on these cultivars; the infected fruits do not develop the unsightly green rings produced by the virus on other yellow-fruited squashes, and consequently, they can still be marketed. 'MultiPik' and other hybrid cultivars with *B* have become commercially important.

Although bush cultivars of *C. pepo* (e.g. most summer squashes) and *C. maxima* (e.g. 'Emerald') have been cultivated for many years, few *C. moschata* cultivars with short internodes, such as 'Burpee Butter Bush', have been available until recently. Researchers in Puerto Rico and Florida are currently selecting for the semi-bush habit in local calabaza landraces.

Buffalo Gourd

Attempts have been made to domesticate buffalo gourd (*C. foetidissima*). This xerophytic perennial is native to the arid deserts of southwestern USA and Mexico. Its ability to withstand drought has attracted interest in its commercialization. Buffalo gourd produces a high yield of seeds, which are rich in protein and oil. The foliage can be used for cattle feed, and the large tuberous roots are a source of carbohydrates. F_1 hybrids of buffalo gourd have been bred by using a gynoecious accession of the species as the female parent and a monoecious inbred line as the male parent. Despite considerable research, however, this species is not yet being grown commercially.

WATERMELON

Introduction

Watermelon, *Citrullus lanatus* (syn. *C. vulgaris* Schrad.) is an important crop in China, Africa, India, USA and other areas with a long, warm growing season. The plants are fairly drought resistant, flourishing on fertile, sandy soils in hot, sunny, dry environments. Worldwide consumption of water-melon fruits and their seeds is greater than that of any other cucurbit (Table 2.1).

Classification

Three species of *Citrullus* are generally recognized: *C. lanatus*, *C. ecirrhosus* Cogn., and *C. colocynthis*. The recently described *C. rehmii* De Winter may represent a valid fourth species (De Winter, 1990). Within *C. lanatus*, domesticated watermelon belongs to var. *lanatus*, whereas wild populations are generally classified as var. *citroides* (Bailey) Mansf. All species in the genus are cross-compatible with each other to varying degrees. *Citrullus lanatus* and *C. ecirrhosus* appear to be more closely related to each other than either is to *C. colocynthis* (Navot and Zamir, 1987). Two additional species closely related to *Citrullus* and once included in the genus are *Praecitrullus fistulosus*, which is cultivated in India and Pakistan for its edible fruits, and *Acanthosicyos naudinianus* (Sond.) C. Jeffrey, a wild species native to southern Africa.

Botany

Watermelon differs from other economically important cucurbits by having pinnatifid leaves. The hairy stems are thin, angular and grooved, bearing branched tendrils. Dwarf watermelon varieties with reduced internode length have been bred, but most commercial cultivars are highly branched vines, measuring up to 10 m in length. The root system is relatively extensive but shallow.

   Watermelon is monoecious. The solitary, light yellow flowers are less showy than those of many other cucurbits. The large, round to oblong or cylindrical fruits measure as long as 60 cm. The 1–4-cm thick rind is hard but not durable; the exocarp is light to dark green, either solid-coloured, striped or marbled. The bland to sweet-tasting flesh is usually red, but may be green, orange, yellow or white in some cultivars or landraces. Watermelons vary considerably in seed colour (e.g. black, brown, red, green or white), shape and size, and seed characteristics can aid in cultivar identification.

Origin and History

All *Citrullus* species originated in Africa, but *C. colocynthis* also grows wild in India. Wild populations of *C. lanatus* var. *citroides*, which are common in central Africa, probably gave rise to domesticated var. *lanatus*. Watermelon has a long history of cultivation in Africa and the Middle East. It has been an important vegetable in Egypt for at least 4000 years. By the tenth century AD, the crop was grown in China and southern Russia. Commercial production of black watermelon seeds has been ongoing in northwestern China for over 200 years. Watermelon was introduced to the New World by the Spanish in the sixteenth century, and quickly became popular with Native Americans. In the USA, it is commercially grown primarily in Florida, Georgia, California and Texas.

Uses

Watermelon fruits make a delicious and refreshing dessert, especially esteemed in hot weather. For centuries, they have served as an important source of water in the Kalahari Desert and other arid areas of Africa. Watermelons are mostly eaten fresh, but in Africa they are also cooked. The rind may be pickled or candied. In the southern states of the former USSR, juice from watermelon fruit is made into a fermented drink or is boiled down to a heavy, sweet syrup.

Watermelon seeds are powdered and baked like bread in India. Roasted seeds are eaten in the Orient and the Middle East, and some Chinese cultivars used for this purpose have been bred to have very large seeds. However, big seeds can be objectionable for watermelons eaten as a dessert, and Japanese breeders have successfully selected for small seed size.

Cultivars and Breeding

Watermelons as large as 100 kg have been reported, but most cultivars produce fruits weighing 4–25 kg. 'Ice box' cultivars are so-called because their fruits, weighing only about 1 kg, fit conveniently in a refrigerator. They have been bred to be sufficiently early to mature in short season areas. The ice box cultivar 'Sugar Baby', introduced by M. Hardin in 1956, produces small, dark green, spherical fruits. It is popular with home gardeners, and has become significant in Asia.

Fusarium wilt resistance, obtained from var. *citroides*, was first bred into 'Conqueror'. Resistant cultivars with improved quality were subsequently developed and became more valuable commercially. 'Klondike R-7', a fusarium resistant cultivar introduced by the University of California at Davis in 1937, was popular in that state for many years.

'Charleston Grey', which produces elongated, light green fruits with red, crisp, sweet flesh of superior quality and flavour, was released in 1954 by the USDA. It soon gained prominence and continues to be a leading cultivar in the USA. 'Crimson Sweet', introduced in 1963, has also become important in commerce because of its high-quality fruits. Both of these cultivars are resistant to fusarium wilt and anthracnose.

'Charleston Grey' and 'Crimson Sweet' are but two of the many cultivars grown in African and Asian countries. Local selections in tropical West Africa include 'Accra', 'Anokye' and 'Volta'. 'Arka Jyoti' and 'Tarmuj' are Indian cultivars, and 'Zhongyu No. 1' is a Chinese landrace.

Seedless watermelon cultivars (e.g. 'Crimson Jewel', 'Genesis') were developed as a result of the discovery by Kihara (1951) that triploid watermelons are virtually without seeds. Tetraploids are produced by the application of colchicine, and can be maintained by selfing or sibbing. They are used as maternal parents in crosses with diploids. The resulting triploid F_1 hybrids, because of their unbalanced chromosome number, are highly sterile. Germinating pollen grains stimulate enlargement of the triploid ovaries. The fruits are not always entirely seedless, but may have small, empty seed coats and an occasional seed.

Because seedless cultivars have reduced amounts of viable pollen, growers are advised to include a normal, diploid cultivar in fields of seedless watermelon in order to improve pollination and fruit set. Typically, one row of the diploid is planted to alternate with two or more rows of the triploid. The diploid and triploid cultivars should be distinguishable, such as one having striped and the other solid-coloured fruits, so that they can be readily separated at harvest. An adequate bee population (one bee to every 100 flowers) is needed for successful pollen transfer or the triploid fruits will be oddly shaped.

Germplasm of triploid cultivars is expensive to produce and may germinate poorly if conditions are less than ideal. Therefore, in order to avoid wasting seeds by poor germination and thinning in the field, growers often start plants in the glasshouse, or in cold frames with bottom heat to promote germination, and then transplant. Some Japanese growers remove the seed coats before planting in order to improve germination. Although the expense of producing seeds is a limiting factor in the use of triploid cultivars, they are grown commercially as well as in home gardens.

In addition to seedlessness and disease resistance, breeders have selected for earliness, high yield, improved flesh characteristics such as greater sugar content, and tough, but flexible rind, the latter to reduce damage during shipment. Compact bush cultivars have been bred and are useful for home gardeners with limited space, but the commercially important cultivars are viny. Similarly, recently developed hybrid diploid and triploid cultivars do not yet compete significantly with the open-pollinated diploids in the marketplace.

Fig. 4.7. Fruit and leaves of citron.

In China, where the seed is valued more than the fruit flesh, scientists continue to select for greater yields of even larger seeds. They are also seeking to improve resistance in their landraces to fusarium wilt, anthracnose, and gummy stem blight (Zhang and Jiang, 1990).

Citron

The fruit rind of preserving melon or citron (*C. lanatus* var. *citroides*) is used to make pickles or conserves, and the fruits are fed to livestock (Fig. 4.7). Citron has white or pale green flesh which is bland to bitter. The large seeds are of various colours, including light green (e.g. 'Colorado Preserving Melon'). In commercial production, over 100 fruits per plant and up to 200 seeds per fruit give seed yields of 500–700 kg ha^{-1}.

In Africa, where the plant is called 'egusi' (a name also applied to *C. colocynthis* and *Cucumeropsis mannii*), the non-bitter seeds are roasted and eaten, or are ground into flour. Oil extracted from the seeds is used for cooking, and the high-protein residue is made into fried seed balls. Local cultivated landraces include 'Akatewa' and 'Nerri' in Ghana and 'Bara' and 'Serewe' in Nigeria.

Citron grows wild in Africa and as an escape elsewhere. It is a weed in watermelon growing areas of North America, causing severe problems in cotton and sorghum fields in Texas, where three distinct citron types exist. Citron crosses readily with watermelon, and their seed production fields should be isolated.

Colocynth

The perennial *C. colocynthis* is cultivated for 'colocynth', a drug produced from the dried pulp of unripe but mature-sized fruits. Although colocynth fruits have bitter, white flesh, the non-bitter seeds are eaten and used for cooking oil in Africa. This species grows wild in northern Africa and southwestern Asia, and exists as an escape from cultivation in other areas, including Australia and southern Europe.

BOTTLE GOURD

Introduction

Lagenaria siceraria (syn. *L. vulgaris* Ser., *L. leucantha* Rusby) is known as bottle gourd, calabash and white-flowered gourd. The genus *Lagenaria* was previously considered to be monotypic, but now six species are recognized, including five wild, perennial species indigenous to Africa: *L. abyssinica* (Hook. f.) C. Jeffrey, *L. breviflora* (Benth.) Roberty, *L. guineensis* (G. Don) C. Jeffrey, *L. rufa* (Gilg) C. Jeffrey, and *L. sphaerica* (Sond.) Naud. Based on morphological differences, bottle gourd cultivars and landraces of African or New World origin are recognized as *L. siceraria* ssp. *siceraria*, and those of Asia are classified as *L. siceraria* ssp. *asiatica* (Kobiakova) Heiser.

Botany

Bottle gourd is a mesophytic annual. The large vines may be more than 10 m long. Stems are angular and tendrils are bifid. The large, cordate leaves are softly pubescent on both surfaces and have a disagreeable odour when bruised. Two bilateral secretory glands occur at the juncture of the long petiole and the lamina.

Bottle gourd is monoecious, but dioecious species also occur in the genus.

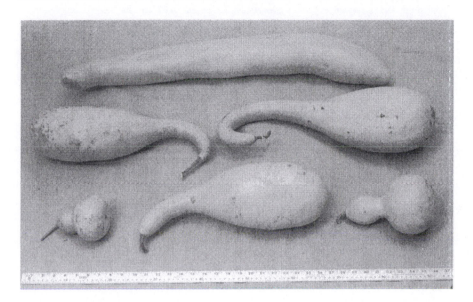

Fig. 4.8. Variation in bottle gourd fruits.

The domesticated cultivars have large, solitary white flowers with long petioles. The night-opening flowers are believed to be pollinated by moths, but may be pollinated by cucumber beetles and other insects as well.

There is great diversity in the size and shape of bottle gourd fruits as a result of thousands of years of human selection in disjunct areas of the globe (Fig. 4.8). Fruit shape ranges from broad and rounded to pyriform, club-shaped, dipper-shaped, curved or cylindrical. Mature round fruits of some cultivars are only 5 cm in diameter, while the narrowly cylindrical types can reach up to 3 m long. Fruit colour is light green mottled with white splotches to white, with the rind turning light brown in very old, dried fruits. Fruits are pubescent when young and glabrous when ripe. The very thick, lignified rind is durable and watertight. The white flesh is moist in immature fruits but papery in very old fruits.

Bottle gourd has large, brown, corky seeds of unusual shape (Fig. 4.9), making them easily recognizable in archaeological excavations. There is considerable genetic diversity in seed dimensions.

Origin and History

Bottle gourd has a longer documented history of use, in both the Old and New Worlds, than any other plant. The firmest archaeological evidence places this species in Peru, Mexico and Florida, USA at least 7000–10,000 years ago and

Fig. 4.9. Bottle gourd seeds, measuring 15–17 mm long.

in Africa over 4000 years ago. It has also been utilized in Asia and the Middle East for millennia. To date, the oldest remains known from Polynesia are about 1000 years old.

Once thought by many to be of South American origin because of the antiquity of its use in Peru, an African origin is now generally accepted for bottle gourd. It migrated to South America, North America, and Asia as a wild or cultivated plant thousands of years ago. The fruits will float in sea water for nearly a year without the seeds losing viability. Consequently, it is believed by some that oceanic drift was responsible for the early worldwide distribution of this species. Dispersal in prehistoric times was undoubtedly aided by people; Islamic nomads are known to carry their valued gourds and seeds with them. Also, bottle gourds used as fishnet floats might have occasionally broken away, carrying seeds to distant shores.

Uses

Immature fruits are cooked like summer squash in India, Italy, China and other countries. They are also added to curries, and the moist flesh is used to make glaze for cakes. In Japan, the flesh of 'Yûgao' is cut into strips and dried in the sun for preservation; the dried fruit shavings are then used as edible wrappings for small amounts of food. Seeds are sometimes made into a vegetable curd, and the softly pubescent young shoots are eaten in China, Italy and elsewhere.

Bottle gourd fruits for consumption need to be non-bitter, but bitter forms can serve as containers for food and water if the bitter factor is leached out by prolonged soaking. Mature fruits have also been used to make musical instruments, birdhouses, ladles and rattles. They are worn as penis sheaths in New Guinea, South America and Africa. In ancient China, small gourds provided cages for crickets kept as pets. In Kenya, long fruits serve as milk pails;

Fig. 4.10. Decorated bottle gourds from South America.

the Masai milk directly from the cow into the gourd, then add cow's blood to the milk and make a fermented drink which is stored in the gourd. Many other uses of bottle gourds are described in the delightful book by Heiser (1979).

Decorating bottle gourds is part of the culture of people of many lands. Independently, natives in Africa, South America and Polynesian islands came to carve and decorate the fruits to make them beautiful as well as utilitarian. Each area has a unique style of embellishing gourds, and the origin of a decorated gourd can often be ascertained from the style of the artist.

The art of gourd carving was highly developed in the Incan Empire. Even today, artisans in Peru create elaborate designs (Fig. 4.10), using only the crudest of tools in the traditional ways of their ancestors. The tip of a *Eucalyptus* stick is singed in fire, and the burnt end used to darken areas of the gourd to various shades of brown. Periodically, the artisan blows on the hot end of the stick to regulate its temperature and achieve the desired colour in the gourd. Light-coloured areas are produced by carving the rind to reveal the paler tissue beneath.

Bottle gourd has long been an important component of indigenous herbal medicine, particularly in Asia. In addition to serving as medicine containers, the mature fruits are prepared as a diuretic, emetic or antipyretic. Leaves, seeds and flowers also have medicinal applications among indigenous peoples throughout the tropics.

Bottle gourds have even been used for currency. In the early 1800s, the ruler of Haiti, Henri Christophe, required his subjects to give his kingdom all of their gourds. The people sorely missed their gourds, since most had been brought with them via the slave trade from Africa, where gourds were considered essential for utensils and many other purposes. Christophe returned the gourds as renumeration for his subject's services, thereby rendering a monetary value to the gourds. Ever since then, monetary coins in Haiti have been called 'gourdes'.

Wild species of *Lagenaria* are utilized in Africa for medicine, fish poison and intoxication. Nigerians include pieces of fruits of *L. breviflora* in preparations to depilate hides and prepare them for tanning.

Cultivars and Breeding

The American Gourd Society, Inc. and other gourd societies boast thousands of gourd enthusiasts from all over the globe, including Australia, Europe and Japan. However, research, breeding and large-scale production of bottle gourd are relatively limited. Nevertheless, international and domestic seed companies continue to offer a variety of cultivars, though not as many as in the past. The most common cultivars sold by seed companies today are those with long, narrow edible fruits, such as the Italian 'Cucuzzi' or 'Longissima', which are harvested when immature and prepared like summer squash. Similar edible cultivars have been bred in Japan and China.

Many cultivars have extensive histories and are often named for their use or appearance. 'Bottle Gourd' or 'Dumb-bell' is an ancient cultivar figured in European herbals; the fruit, which is constricted in the centre and bulbous at the blossom and peduncle ends, has long been used as a storage container. 'African Kettle' and 'Bushel Gourd' produce broad, rounded fruits measuring around 50 cm in diameter; these are sometimes fashioned into percussion musical instruments. 'Birdhouse' is actually a collection of heirloom selections with fruits suitable as nesting houses for small birds. 'Dolphin', also called 'Maranka', has an unusual fruit with contorted rind ridges (Fig. 4.11). A few bottle gourd landraces in Africa produce warted fruits, but these are not generally commercially available.

As scientists take a greater interest in the increasingly popular bottle gourd, some of their surveys have naturally focused on disease resistance. For example, 'Cow Leg', one of the most common cultivars grown in Taiwan, was recently shown to be resistant to several viruses, including CMV, squash mosaic virus (SqMV), WMV and ZYMV (Provvidenti, 1995).

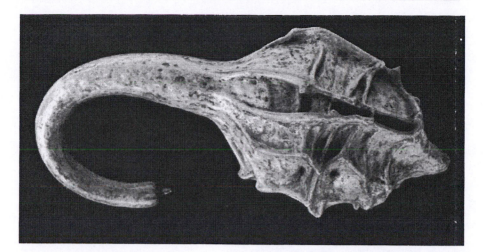

Fig. 4.11. 'Maranka', an unusual cultivar of bottle gourd. This fruit is about 30 cm long.

Culture

Bottle gourd is chiefly suited to sunny, semi-dry, low-elevation areas, but can be grown in wetter tropical climates as long as soils are well-drained. Temperatures of 18–30°C are acceptable, and the plants may be grown in glasshouses in temperate countries. Trellising is necessary if long, thin fruits are to be straight, but is not recommended for cultivars grown for very large, heavy mature fruits (e.g. 'African Kettle'). Periodic irrigation may be required during dry periods, and fertilizers for nutrient-poor soils.

Fruit shape can be modified with twine or moulds if begun at the earliest stages of development. In fact, over time, a long gourd can be tied into a knot by the patient and persistent gardener.

Immature fruits are typically harvested 60–90 days after sowing. Fruits of cultivars grown for mature bottle gourds must be left on the vine at least another month, and often longer, before they reach maturity. Average yields are about 25 t ha^{-1} at a density of 6000 plants per hectare, and yields can be increased with growth regulator applications (Tindall, 1983).

Diseases and pests are relatively few, although viruses and powdery mildew are sometimes a problem. Anthracnose has become a recent concern in bottle gourd fields in India. Young fruits can be covered with mosquito netting to keep insects from laying eggs in the rind.

LOOFAH

Introduction

Fruits of *Luffa* somewhat resemble cucumbers, and Linnaeus included loofah species in the genus *Cucumis*; however, these genera are unrelated, belonging to different tribes (Table 1.1). The domesticated species, *L. acutangula* and *L. cylindrica* (syn. *L. aegyptiaca* Mill.), and two wild species have their origins in the Old World, whereas the remaining three wild species are indigenous to South America. The South American species are the only members of the tribe Benincaseae known to be native to the New World. Heiser and Schilling (1990) speculated that the genus originated in Gondwanaland and spread before the continents drifted apart, or that oceanic dispersal occurred after landmass separation. Given that the fruits and seeds can float, the latter scenario seems plausible.

Luffa *cylindrica* (called smooth loofah, dishcloth gourd or sponge gourd) consists of wild populations, ranging from southern central Asia to north-eastern Australia and the South Pacific, and the domesticated variety, which is cultivated in Asia, Africa, various tropical American countries and the Caribbean. *Luffa acutangula* (angled loofah, silk gourd, ribbed gourd, Chinese okra) contains three varieties: var. *acutangula*, which is grown in southeastern Asia and other tropical areas, but to a lesser extent than smooth loofah; var. *amara* (Roxb.) C. B. Clarke, a wild or feral form confined to India; and var. *forskalii* (Harms) Heiser & Schilling of Yemen, which is probably a feral form. The area of original domestication for smooth loofah is unknown; India is probably the centre of origin for domesticated angled loofah.

Botany

Loofah vines are large, sometimes more than 10 m long. The slightly hairy stem is five-angled. Smooth loofah has large, shallowly to deeply lobed leaves, which are mottled with white or silvery areas between the veins when young. Those of angled loofah are shallowly lobed to merely angled and lack the mottling. Tendrils are usually 3–5-branched (Fig. 1.1).

Most loofah species are monoecious, but *L. echinata* Roxb. is dioecious. The yellow flowers are relatively large and conspicuous. Female flowers are solitary, often occurring in the same leaf axil as the long-stalked raceme of male flowers. The five stamens are free in smooth loofah and fused to appear as three stamens in angled loofah. Flowers of smooth and angled loofah open in the early morning and late afternoon, respectively. Loofah species have glands that secrete considerable nectar, attracting ants and a variety of pollinating insects. Extrafloral nectaries occur at the base of pedicels, in some leaf axils and on the lower surface of pistillate calyx lobes (Fig. 4.12).

Fig. 4.12. Young fruit of smooth loofah, with close-up of ants feeding at the nectaries on the persistent sepals.

Wild species and varieties of *Luffa* have small, bitter fruits. The cylindrical to club-shaped fruits of the domesticated varieties are mostly non-bitter and measure 15–60 cm in length or longer. Fruits of angled loofah differ from those of smooth loofah by having ten pronounced longitudinal ribs. Fruits of both species are bright green when immature, turning yellow to brown when mature. Mature fruits of all loofah species dehisce at the blossom end, enabling the black, brown, or rarely white seeds to disperse from the dry, fibrous endocarp. Seeds of the domesticated varieties are compressed, ovoid and about 12 mm long.

Uses

Immature fruits of the domesticated varieties, especially angled loofah, are harvested when about 10 cm long, and then are boiled, peeled or intact, and used in curries. Typically, only non-bitter forms are eaten as vegetables, prepared in ways similar to summer squash or sometimes eaten fresh like cucumber. In China, shavings of loofah are used as a garnish. The flavour of the fruit is bland, not unlike cucumber, but with the texture of zucchini. Leaves are also edible and used as greens, and mature seeds are sometimes roasted.

Various parts of wild and domesticated loofah species have been employed

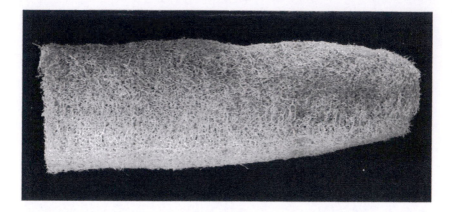

Fig. 4.13. Fibrous, spongy interior of smooth loofah.

in herbal medicine, usually as a purgative, in Asian, African and South American cultures. The Japanese extract stem sap for a respiratory medication and to make into various commercial toiletry items with names like 'Eau de Hechima' and 'Dentrifice de Hechima' (Porterfield, 1955).

Smooth loofah is sometimes called sponge gourd because its mature fibrous endocarp is often used for that purpose (Fig. 4.13). These 'environmentally friendly' sponges have become very popular in the USA, which imports millions of them each year from Asia. The fibrous endocarp has also been used for sandals, insulation, soles of slippers, padding for saddles, stuffing for pillows, and many other purposes. During the early 1900s, loofah sponges were utilized primarily as filters, particularly as oil filters in American, British and German steamships. Most of those sponges were produced in Japan and Brazil.

Today, Korea, China, Guatemala and Colombia are among the major producers of loofah sponges. The crop is also grown commercially in southern USA. Seeds of both domesticated species are offered by many international seed companies.

Cultivars and Breeding

Commonly grown cultivars of angled loofah in India are 'Pusa Nasdar', 'Tori' and the small-fruited, hermaphroditic 'Satputia'. A cross between 'Satputia' and 'Tori' may produce five times the yield of either (Chakravarty, 1990).

The greatest need in breeding smooth loofah is for standardization of fruit characteristics. As with angled loofah, distinct cultivars and landraces have evolved in different countries, but they have not been adequately investigated and classified. In China, there is ample diversity in smooth loofah types. Grown commercially in the Yangtze River regions are long-fruited cultivars, e.g. 'Xian-si-kua' and 'Hu-Lu-qing', which produce fruits up to 150 cm in length.

In the past 20 years, research relevant to commercial loofah sponge production has been conducted in tropical West Africa and India. Popularity of sponges in the USA has led to recent investigations on production practices for temperate regions of that country (Davis, 1994).

Culture

Loofah plants grow best on fertile, mildly acidic to neutral, well-drained soils during the dry season in sunny, hot, humid areas. They are frost-sensitive and somewhat drought tolerant. Angled loofah is less tolerant of wet growing conditions and more resistant to diseases and pests than is smooth loofah. Angled loofah can be cultivated at elevations of up to 500 m, and smooth loofah to 1000 m. Although smooth loofah is adapted to short daylengths, angled loofah is photoperiod insensitive.

Under glasshouse culture, smooth loofah plants require periodic feeding with nitrogen, temperatures of 25–28°C, and spraying for pest and mildew control. For field-grown crops, seeds or transplanted seedlings should be planted in hills at the base of a sturdy trellis. Within-row plant spacing and pruning can affect fruit size and sponge quality in smooth loofah (see Chapter 5 for more details). The earliest female flowers are often culled and only a limited number of fruits (<20) allowed to mature on the vine. Immature fruits can be harvested 40–80 days after planting; harvesting of mature fruits begins about 3 months later. In Japan, mature fruit yields for smooth loofah reach up to $50 \, t \, ha^{-1}$. Yields of sponge-quality fruits have been reported as high as 77,000, 60,000 and 48,000 fruits per hectare in the USA, Japan and Korea, respectively.

BITTER MELON AND RELATED SPECIES

Introduction

The genus *Momordica* contains about 45 species native to the paleotropics, chiefly Africa. The name for the genus is derived from the Latin word 'mordicus', meaning bitten; it refers to the fact that the jagged edges of the seed look like bite marks. At least six species are under cultivation, and others are variously used. Two species, *M. charantia* and *M. balsamina*, have escaped from cultivation and become naturalized in the neotropics. Weedy *M. charantia* is particularly widespread and a nuisance in agricultural areas such as citrus groves in Florida. These plants can flower and fruit throughout the year as long as temperatures remain high.

Bitter Melon

Botany

Momordica charantia (bitter melon, balsam pear) vines, growing to 4 m, are slender and glabrous to slightly hairy. Although the tender stems eventually die back, new shoots arise from the perennial stem bases attached to the large succulent roots. The highly lobed leaves are borne on long petioles. Tendrils are unbranched, reaching up to 20 cm in length. Plants are usually monoecious, rarely hermaphroditic, with solitary yellow flowers in the leaf axils. The light green fruit has ten longitudinal ridges and a very irregular surface studded with protuberances (Fig. 2.4). It is spindle- or pear-shaped, measuring up to 25 cm long in the domesticated variety and about 5 cm long in wild populations (var. *abbreviata* Ser.). At maturity, the fruit turns orange and splits open, usually starting at the blossom end, to reveal the bright red, fleshy arils surrounding the seeds. Birds relish the sweet arils and are seed dispersers. The sculptured, ovate seeds have undulate margins and are various shades of brown, sometimes with black markings, particularly in wild populations.

Origin and history

Bitter melon is a very popular vegetable in India, the Philippines, Malaysia, China, Australia, Africa, the Middle East, Latin America and the Caribbean. This species was domesticated in Asia, possibly in eastern India or southern China, and wild or naturalized populations occur throughout the paleo-tropics. Its introduction to the New World apparently took place during the slave trade; it subsequently escaped from cultivation, becoming a common weed in tropical and subtropical regions from Brazil to southeastern USA.

Uses

Bitter melon is an ornamental plant, with its delicate foliage, bright yellow flowers and colourful fruits of unusual shape. Adding to its appeal, the flowers have a strong vanilla-like scent in the morning. It has been grown as a decorative plant in British glasshouses for more than a century. The vines are often trained on trellises. A weight is sometimes tied by string to the developing ovary in order to produce a very straight fruit.

Because mature fruits are very bitter, fruits for consumption are usually harvested before maturity, when they are just turning orange but still firm.

This bitterness is due to momordicosides, which are glycosides of tetracyclic triterpenoids with a cucurbitane skeleton. Immature fruits are less bitter and have a tangy flavour that adds spice to oriental cuisine. To reduce bitterness further, the fruit is steeped in salt water before cooking. Then, the fruit is boiled, curried, or sliced and fried. It may also be pickled or dried for later consumption. The fruits are rich in iron, calcium, phosphorus and vitamins.

Young shoots, leaves and flowers are steamed and eaten as potherbs or used as seasonings. Children sometimes eat the sweet arils, but seed ingestion should be avoided because the seeds are toxic, containing alkaloids. Cases have been reported of people becoming ill from eating bitter melon fruits or vines, and ingestion of the seeds can be fatal.

Alkaloids, triterpenes, a saponin-like substance and other biodynamic compounds are present in various parts of the plant. Consequently, fruits are used to make lather, the plant can be ground into an insecticide or a poison, and the vine, leaves, roots, fruits, seeds and flowers are employed to treat a plethora of ailments in China, Africa and elsewhere (Morton, 1967). During the past decade, medical researchers have studied the antidiabetic properties of the plant, which function by increasing carbohydrate utilization; scientists have isolated a hypoglycemic principle called charantin. Bitter melon is also being investigated as an agent that inhibits multiplication of the HIV virus.

Cultivars and breeding

There has been little research to define and standardize cultivars, most of which have their origins in regional landraces. A variety of types with different cultivar names (e.g. 'Chinese Bitter Melon', 'Thailand Bitter Melon' and 'Bitter Gourd Taiwan Large') are offered by seed companies. In China, there are apparently three horticultural groups of cultivars. Over the past century, at least two types of cultivars have been grown in northern India, and nine types in southern India. These cultivars (e.g. 'Arka Harit', 'Karela', 'Jethuya' and 'Asadhi') differ in their growth habit, maturation period and various fruit characteristics, including size, shape, colour and surface texture. Today, Indians cultivate primarily two varieties of M. charantia: var. charantia, which produces large, fusiform fruits, and var. muricata (Willd.) Chakravarty, with smaller, rounder fruits. Indian plant breeders have crossed some of their cultivars, producing vigorous hybrids with large fruits and thick flesh. Currently, they are experimenting with interspecfic hybridization in an attempt to transfer desirable attributes, like lack of fruit bitterness, from kaksa and other related species to bitter melon.

Culture

Bitter melon grows well in hot, humid areas at elevations up to 500 m. Rich soils with high water-retaining properties are preferable. Irrigation may be needed once or twice a week during dry weather. Yields can be improved with applications of around 100, 50 and 50 kg ha^{-1} of N, P and K, respectively. Plants should be trained on trellises. Flowering begins about 5 weeks after sowing. Depending on the cultivar, young fruits become ready to harvest 1–3 weeks later, while still firm and weighing 80–120 g. Yields are 8–15 t ha^{-1} from plantings of a density of 40,000 plants per hectare (Tindall, 1983). For seed production, fruits should not be picked until they have turned orange and begun to dehisce.

Diseases and pests of the foliage are relatively few, apparently because of the various toxic compounds in the plant. Bitter melon is susceptible to WMV and other cucurbit viruses, though, and fruit rot can occur during the rainy season. Powdery mildew is controlled with sulphur dust or other chemicals. When developing fruits are bagged as protection from fruit flies, the fruits turn white.

Related Species

Momordica balsamina

Momordica balsamina, known as balsam apple, is sometimes confused with wild plants of bitter melon, which have similarly sized fruits. However, the mature, red fruit of balsam apple is usually beaked and not as warted as bitter melon. Balsam apple also differs from bitter melon by having relatively smooth seeds, very slender, short stems, leaves that are less deeply lobed and shorter staminate pedicels.

Balsam apple fruits are sometimes eaten. More often, they are utilized in medical preparations, in ways similar to bitter melon. Like bitter melon, this species is widely adventive throughout the tropics.

Momordica cochinchinensis

Momordica cochinchinensis is known as Cochinchin gourd, spiny bitter cucumber, and sweet gourd of Asam. It may have originated in India, and is cultivated and naturalized throughout southern Asia.

Cochinchin gourd is generally described as dioecious, although there are reports from India of monoecy. Dioecious plants are propagated by cuttings to enable the grower to have a high proportion of female plants. Cochinchin gourd is also seed propagated. Germination is hypogeous.

Vines are vigorous, with large leaves and thick, woody stems. The large, solitary flowers are cream-coloured to pale yellow, with black areas at the bases of the petals. The 10–20-cm long, ovoid fruits are covered with raised conical points and are bright orange or red when mature. Embedded in the orange-red pulp are large (ca 4 cm across), flattened, black seeds that are sculptured with jagged edges (the second-largest seeds shown in Fig. 1.5, page 13).

After removing the spiny skin, immature fruits are cooked as a vegetable or prepared in curries. Young leaves, flowers and seeds are also edible. Oil can be extracted from the seed and used for illumination. The roots, which contain a triterpenoid saponin, produce froth when rubbed in water, serving as a substitute for soap. Various parts of the plant have medicinal uses in China, India and elsewhere in southern Asia.

Cochinchin gourd is usually grown during the rainy season in the hot, humid tropics; otherwise, ample irrigation may be necessary. Initial application of N:P:K is often supplemented with a sidedressing of nitrogen when fruits begin to set. Immature fruits are harvested at 1–2 weeks old. Diseases are few, but fruit flies can cause fruit damage.

Momordica dioica

Momordica dioica (kaksa) is native to southern tropical Asia, growing wild from India to China. It is a dioecious perennial with solitary, yellow flowers. The small (3–8 cm long) fruits are sweet, non-bitter and have soft spines. Seeds are pale yellow.

Kaksa is cultivated in India, where the immature fruits are used in curries. Young shoots and leaves are cooked as a vegetable. The succulent roots, which are larger in female plants, are also eaten. Because they are astringent and contain traces of alkaloids, the roots are used medicinally.

This species will not naturally cross with bitter melon, which differs in sex expression, chromosome number and karyotype. However, Indian researchers have recently hybridized the two species and are beginning to select for desirable gene combinations.

Momordica angustisepala and M. cymbalaria

Momordica angustisepala (sponge plant), a native of Africa, is sometimes cultivated in Ghana. The thick stems are pounded and worked to yield white, absorbent fibres that are used in washing. *Momordica cymbalaria* (syn. *M. tuberosa* (Roxb.) Cogn.) ranges from northeastern Africa to India. It is cultivated occasionally in India, where fruits provide food and medication.

WAX GOURD

Introduction

Benincasa is named for Count Benincasa, an Italian patron of botany. The only species in the genus is *Benincasa hispida*, which is known as wax gourd, winter melon, white pumpkin, ash gourd, Chinese preserving melon, Chinese squash and other names. The name wax gourd refers to the thick, waxy cuticle that typically develops on mature fruits. High temperature favours formation of the waxy bloom. The specific epithet previously used, *cerifera*, means wax-bearing. The specific epithet now recognized, *hispida*, refers to the hirsute pubescence on the foliage and immature fruit.

Wax gourd is an important vegetable in China, India, the Philippines and elsewhere in Asia. It is also grown in Latin America and the Caribbean, usually by immigrants of Chinese descent. This species has been cultivated in China for more than 2300 years, and is believed to have originated in southeastern Asia. Small-fruited wild populations, classified as *B. hispida* var. *pruriens* (Parkinson) Whistler, occur on several islands in the South Pacific (Whistler, 1990) and reportedly in Australia, Indonesia, Japan and southern China as well.

Botany

Wax gourd, with its showy yellow flowers, large leaves, long vines and huge fruits, bears some resemblance to *Cucurbita*. Early taxonomists classified it as *Cucurbita hispida* Thunb. The three stamens of *Benincasa* flowers are separate, however, whereas those of *Cucurbita* are more or less united. *Benincasa* has male flowers with long pedicels, but the female flowers are almost sessile, not borne on long pedicels like those of *Cucurbita*. *Benincasa* has a probract at the base of each leaf petiole, but *Cucurbita* does not. The two genera, which occur in distinct tribes (Table 1.1), also differ in many other characteristics, including number and arrangement of vascular bundles, biogeography and chromosome number.

The hairy vines of wax gourd are large and spreading. Each tendril is 2- or 3-branched, occurring opposite the probract at the leaf axis. The large, hairy leaves are lobed and have long petioles; they emit an unpleasant odour when bruised. Wax gourd is monoecious, with solitary male and female flowers. Above the shallow hypanthium, the petals are almost completely separate and widely spread (Fig. 6.2, page 149). Sepal lobes are foliaceous.

Mature fruits are heavy, weighing up to 40 kg, depending on the cultivar. Some resemble a watermelon with flattened ends, and may be over 1 m long. In other cultivars, the fruits are small and cylindrical, or globose (Fig. 4.14). Rind colour is green with light-coloured speckles. Immature fruits are pubescent; mature fruits may be glabrous or pubescent, some having a dense pelt of minute hairs. The hard, dry rind of the mature fruit is usually covered with a white waxy bloom. Fruit flesh is white, crisp and juicy. The buff-coloured seeds are flat, with margins that are ridged in some cultivars and smooth in others.

Mature fruits are called winter melons because they can be stored for as

Fig. 4.14. One small, wild and two large, domesticated fruits of wax gourd. The small fruit measures about 5.5 cm in diameter. Abrasions in the waxy coating reveal the underlying green rind. The fruits are also covered with a mildly prickly pubescence, best seen on top of the fruit on the right.

long as a year. The waxy coating serves to keep moisture in and insects and microorganisms out. They have a very high moisture content (ca 96%) and are low in calories and carbohydrates (Morton, 1971).

Uses

Both mature and immature fruits of the bland-tasting but easily digested wax gourd are consumed. They may be eaten raw, like cucumbers, but more often are cooked or pickled. Immature fruits are prepared like summer squash, whereas chunks of the mature fruit are used to make soup, which is sometimes canned. In China, mature wax gourds are sold by the slice. Occasionally, the entire mature fruit is steamed, often stuffed with lotus seeds, vegetables, meat or other ingredients. In India, the fruit is boiled in sugar syrup to make a confectionery known as 'pithe'. A wax gourd is sometimes given to the bride and groom at the wedding feast in India as a token of good luck.

The Chinese use wax gourds not only as soup ingredients but also as elegant serving bowls for the soup. For ceremonial occasions and formal

Fig. 4.15. Carved wax gourd fruit used as a soup serving bowl in China.

banquets, the interior of the fruit is scooped out and filled with soup. The exterior is exquisitely inscribed into intricate portrayals of birds, dragons or other designs by carving away parts of the green rind to reveal the white tissue below (Fig. 4.15).

Young leaves, vine tips and flower buds are boiled and eaten as greens. Seeds are consumed fried, but usually as a medication instead of a food. In Malaysia, the cooling juice of the plant is rubbed on bruises. The Chinese apply rind ashes to wounds. Seeds, fruits, leaves and roots enter into various other medications throughout southern Asia.

The fruit wax, which will develop even after the fruit is harvested, is sometimes used to make candles. After the wax is scraped off the fruit, still more may form. The wax has also been used by Malaysians as a vehicle for carrying poison for homicide.

Cultivars

In a way similar to loofah, landraces and cultivars of wax gourd have evolved locally and vary among geographic regions. Seed companies offer several cultivars distinguished by their fruit characteristics. Those with names like 'Fuzzy Squash' or 'Chinese Fuzzy Gourd' belong to *B. hispida* var. *chieh-qua* How. The mature fruit of this variety is small (20–25 cm long), cylindrical, hairy and with little or no waxy bloom. Fruits of these and other small-fruited cultivars (e.g. 'Beijing Yi-chuan-lin') are usually eaten while immature.

There are many large-fruited cultivars in China, including 'Guang-dung-Quig-pi', 'Hui-pi Dong-gua' and 'Beijing Di-dong-gua'. These fruits are consumed when mature. In India, cultivars with large, round to oblong fruits and a thick waxy coating predominate and were the earliest wax gourd introductions to Europe.

Culture

Wax gourd grows best in hot (>25°C), sunny, moderately dry areas of the tropics at elevations below 1500 m. However, because the plants grow very quickly, they can also be cultivated in temperate areas with shorter growing seasons. Where temperatures remain above freezing, two crops per year can be produced.

Garden plants are grown over dwellings, on bamboo frames or up on to trees in Asia. Commercial production is either on the ground or on sturdy trellises. Fertile (or fertilized), well-drained soils of pH 5.5–6.4 are preferred. Ammonium sulphate (ca 20 t ha^{-1}) or other source of nitrogen can be added as a sidedressing every 2 weeks until flowering. Wax gourd is relatively drought tolerant, but may require irrigation after 1–2 weeks without precipitation. Hives of honeybees ensure pollination. Only one fruit should be allowed to mature on each lateral branch. Immature fruits are harvested about a week after anthesis (60–80 days from sowing), and mature fruits 2–3 months later, depending on the cultivar. Large fruits on trellises may need additional support so that the vines do not break. Approximately 2 kg of seeds are needed to plant 8000 plants in one trellised hectare, giving a yield of up to 20 tonnes of fruit (Tindall, 1983).

There are no serious diseases and relatively few pest problems. In India, common pests are fruit flies, beetles, aphids and jassids. Because of its resistance to soilborne diseases, wax gourd is sometimes used as rootstock for grafted melon.

CHAYOTE

Introduction

Sechium edule is known as chayote, cho cho, mirliton and vegetable pear. The word chayote is derived from 'chayotli', its name in the ancient Aztec civilization of Mexico. *Sechium* was formerly regarded as monotypic, having only a single species in the genus, but now eight species are recognized.

Chayote is native to Mexico and Guatemala, where it has been grown for

centuries and remains popular today. Also in this region are wild populations
of this species and of its closest relative, *S. compositum* (J.D. Sm.) C. Jeffrey,
either of which could represent the wild progenitor of domesticated chayote
(Newstrom, 1991). From commercial fields in Mexico, Costa Rica, Brazil,
Puerto Rico and Italy, chayote is shipped to markets in the USA, Canada and
Europe. This crop also has become an important vegetable in many tropical
and subtropical areas of Asia, Africa, Australia and South America.

Botany

Chayote is a vigorous grower, with vines reaching up to 20 m in length. The
angled, furrowed stems and large, branched tendrils are glabrous or nearly so.
Leaves, resembling those of cucumber, are broadly ovate to triangular and
angled or slightly lobed.

Chayote is a perennial, developing large succulent roots over several
years. In temperate regions, the vines are killed by frost in the autumn, but
if the winter is not severe, growth resumes in the spring from adventitious
buds at the crown of the rootstalk.

Chayote is monoecious, with a single pistillate flower and a cluster of
staminate flowers at the same node. The flowers are small, with green or white
petals, and less conspicuous than those of most other cucurbits. Inside and at
the base of the saucer-shaped hypanthium, each flower has ten pouch-like
nectaries, which attract a variety of pollinators.

A short photoperiod is required for chayote to flower. Since it may not
flower until the autumn in temperate regions, a long growing season may be
required. This crop is generally grown only in subtropical and tropical areas,
where it may flower all year.

Chayote fruits are white to dark green, pear-shaped to round, 7–20 cm
long, often wrinkled and sometimes with a few spines. There is considerable
genetic variation in fruits with respect to size, shape, colour and spines. The
single-seeded, fleshy fruit (Fig. 4.16) is viviparous, the developing seedling
feeding off of the pericarp (Fig. 4.17). A fruit may even sprout while still on
the vine. The white seed is large (up to 8 cm long) and flattened.

Uses

The immature fruit is cooked similar to summer squash and generally has a
mild flavour. Chayotes can be stored for several weeks at cold temperature
(10–15°C) or be preserved by pickling. In industry, the bland-tasting fruits are
used as filler for tomato ketchup and baby food. Shoot tips, young leaves and
tendrils are also boiled and eaten. The roots, a good source of easily digestible
starch, are harvested in Mexico and boiled, roasted, or candied in sugar and

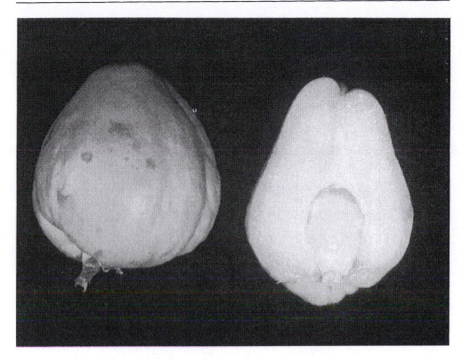

Fig. 4.16. Chayote fruit with fleshy seed exposed.

fried. The leaves, possessing cardiovascular modifying properties, have been used in herbal medicine for hypertension. Fruits and roots are considered diuretic, and seeds are taken for intestinal problems. All parts of the plant make good fodder. In the early 1900s, silvery white straw was extracted from the stems and made into hats and baskets, which were sold in France.

Cultivars and Breeding

There is ample genetic variation in chayote landraces and wild populations in Mexico and Central America that could be used for crop improvement. However, only two commercial types are being exported from Mexico, Costa Rica and Puerto Rico. The dominant type is a bland-tasting, obovoid, light green, smooth fruit that is actually intended for food industry use, but is also sold as a vegetable. The second type is a small, round, white, smooth fruit. Two cultivars currently grown in Florida are 'Florida Green' and 'Monticello White'. Newstrom (1990) suggests that selections from other native landraces could produce a more palatable vegetable for export as well as providing the disease resistance that is lacking in the narrow gene base of commercial

Fig. 4.17. Vivipary in chayote.

cultivars. Costa Rican researchers are in the process of developing such a
breeding programme.

Culture

In the tropics, chayote can be cultivated at elevations of up to 2000 m. It
grows best in areas of fertile soils, high humidity, ample rainfall and mild
temperatures with relatively cool nights. Fertilizer should be added before
planting and once or twice more before fruiting. Frequent irrigation is
required if precipitation is lacking.

For propagation, the entire fruit is often planted at an angle, with the
enlarged end down. Or, the fruit flesh can be removed and consumed, and the
bare embryo planted. Chayote is also grown from cuttings to perpetuate
desirable plants and avoid the genetic diversity of plants grown from seed.

In India, the long vines are trained to grow on horizontal trellises from which the fruits hang down. Plants on a trellis or other support should be widely spaced, at least 3 m apart, because of their large size. They can be grown at a density of up to 1500 plants per hectare.

Flowering occurs during daylengths of around 12.5 h or less. Fruits require pollinators, although parthenocarpic fruits can be obtained with the application of gibberellins. Harvesting of immature chayotes begins about 100 days after sowing. In some areas, harvest is almost year-round; a single plant can produce hundreds of fruits, although 75–100 fruits per plant is more typical. Although a planting can yield fruits for several years, commercial production plants are usually kept for less than 3 years because of attack by nematodes, fungi, viruses and spider mites. A disease plaguing farms in central Mexico causes the petals to become foliaceous.

In Mexico and Guatemala, starchy tubers are harvested from 2-year-old plants. In China, shoots about 20 cm long are clipped and sold in markets as edible greens.

OTHER CUCURBITS

In addition to those already discussed in this chapter, there are many other cucurbits cultivated for food and other uses (Table 1.1). Some of the more important species are:

Snake Gourd and Pointed Gourd

There are about 40 species of *Trichosanthes*, many of which are grown in China for medicinal purposes. The most widely grown species, *T. cucumerina* var. *anguina* (L.) Haines (snake gourd), is cultivated in humid subtropical and tropical countries of Australasia, Latin America and Africa. It is probably native to and was originally domesticated in India.

The long vines of snake gourd have palmately lobed, densely pubescent leaves and branched tendrils. The beautiful white, fringed flowers are fragrant and open at night. Male racemes and solitary female flowers occur on the same plant. Snake gourd gets its name from its very long (up to 150 cm), slender, serpent-like fruit. Immature fruits are green with white stripes, turning orange or red at maturity. The white, fibrous flesh contains brown seeds.

As it ages, the fruit of some cultivars may become bitter; hence, only immature fruits are typically consumed, usually boiled or in curries. Stem tips and leaves are also eaten, and roots and seeds have medicinal applications.

Because the plants are intolerant of dry soil conditions, planting typically takes place at the beginning of the rainy season. Stable temperatures above

25°C and short daylengths promote growth. Vines are trained on trellises or other supports. A stone or similar weight is sometimes tied to the blossom end of the fruit to cause it to grow straight. Young fruits are harvested 2–3 months after planting, when 30–60 cm long. They can be stored up to 2 weeks at 16°C and 85–90% relative humidity.

Pointed gourd (*T. dioica*) differs from snake gourd by being dioecious and having smaller fruits (ca 12 cm long). Pointed gourd is a perennial vine, with simple or bifid tendrils and solitary flowers. It is cultivated primarily in India, where a single planting is maintained to produce two annual crops. Propagation is usually by stem or root cuttings so that the grower can arrange to have mostly female plants. Parthenocarpy can be induced with growth regulators. Immature fruits are boiled, pickled or used in curries, and the young shoots and leaves are eaten.

Fluted Pumpkin and Oyster Nut

Telfairia occidentalis is called fluted pumpkin because of the ten prominent ribs on the fruit. This West African native is cultivated in Nigeria, Ghana, Sierra Leone and other tropical African countries. Vines grow to 15 m long, with foliage that is tinted purple. Fluted pumpkin is a dioecious perennial, although Akoroda *et al.* (1990) discovered a monoecious plant. The white, fringed flowers occur in long racemes at the ends of branches on male plants and are solitary on female plants. The pale green, ellipsoid fruits, which have a white waxy bloom when mature, grow to 1 m long. The red, ovate seeds are up to 5 cm in diameter.

This species is grown primarily for the nutritious stem tips and leaves, which are boiled and eaten. Male flowers are sometimes added to the cooked greens. The large seeds, which have an almond flavour, are prepared for consumption in various ways.

Fluted pumpkin is grown from seeds or vine cuttings. Because the seed produces multiple shoots upon germination, the large cotyledons of a young seedling can be divided and sectioned so that each division (up to four) is complete with a shoot, some roots, and part of a cotyledon.

The vines, which are trained on to supports or grown along the ground, respond well to fertilizer (see 'Fertilizer', Chapter 5, page 135 for more details) and irrigation. Although fluted pumpkin plants are best raised during the rainy season (April–October) in West Africa, irrigation can be used to extend the growing period into the dry season. When the plants are grown primarily for edible greens, young fruits should be removed as they appear. Female plants are hardier and more productive as a leafy vegetable than male plants. Frequent harvesting of young shoots on mature plants encourages branching. Fifteen harvests can be made during a 5-month harvest period, which begins about 3 months after sowing. When seeds for consumption are desired, the fruits should be picked just before maturity.

Oyster nut (*T. pedata*) is another dioecious perennial native to and cultivated in Africa. Vines grow to 35 m long. Leaves have five to seven leaflets. Tendrils are bifid, with one branch longer than the other. Flowers are purplish-pink with fringed margins. The green, ridged fruits have a constricted neck and a distinct rim at the base. They can reach 1 m long and weigh 13 kg. The fruit splits at maturity while still on the vine, releasing as many as 200 large, brown seeds. The seed faintly resembles an oyster, hence, the common name of the species.

Although the vine tips and leaves are cooked and eaten, this species is grown primarily for its seeds. The seed coats are bitter and must be removed before eating the seeds, which are consumed raw, roasted, pickled, or in soup or sweets. Edible oil can be extracted from the seeds also.

Ivy Gourd

Coccinia grandis occurs wild and cultivated from Africa to southeastern Asia. Ivy gourd is also grown and has escaped in Latin America, Australia and other tropical areas.

This species is a dioecious perennial, with glabrous stems growing to 20 m in length. Flowers are white in Indian cultivars, but apparently white or yellow in African landraces (Jeffrey, 1967). The fleshy, ellipsoid fruits measure 6–10 cm long. They are green, often with white stripes, when young, turning scarlet at maturity. Mature fruits are consumed raw, cooked or candied. Leaves, young shoots and immature fruits are cooked as vegetables. Various plant parts are used medicinally. The plant is sometimes grown as an ornamental on arbors.

Ivy gourd is usually propagated by cuttings in order to have about ten times as many female as male plants in the field. Stem cuttings are placed 1–2 m apart beneath a trellis. Irrigation may be required and the application of nutrients desirable. Fruits of some landraces can develop parthenocarpically. A single plant may bear over 300 fruits in a season.

Stuffing Cucumber

Cyclanthera pedata var. *edulis* Schrad. is native to the Americas, where it often occurs as an escape. Stuffing cucumber has also long been cultivated in Asia, sometimes at elevations of up to 2000 m because of its cold tolerance.

The glabrous vines have a strong odour when crushed, somewhat like a cucumber. Leaves are large and deeply palmately lobed. The species is monoecious, with staminate flowers in racemes and solitary pistillate flowers. The greenish or white flowers are relatively small for a cucurbit. The puffy, partially hollow fruit is like a bladder; it is yellowish-green, measures 5–15 cm long, and has soft spines, a tapered neck and black seeds.

The foliage is eaten fresh or cooked. Fruits are eaten raw as a substitute for cucumber or cooked, often after removing the seeds and stuffing the fruit with various ingredients. Seeds are edible as well. Fruits can be harvested throughout most of the growing season.

Casabanana

Sicana odorifera, one of three species in its genus (Lira-Saade, 1995), derives its specific epithet from its very fragrant flowers and fruits. Casabanana, which is probably native to Peru or Brazil, is cultivated in Central and South America and the Caribbean.

This species is a monoecious perennial with stems measuring 15 m long or more. The yellow flowers are solitary, large and relatively tough, persisting for several days after anthesis, during which time they turn pinkish-orange. The cylindrical fruits are yellow, red, dark green or dark purple, and measure 30–60 cm long. The juicy, yellow flesh smells sweet, but is tough.

Casabanana is grown as a fruit vegetable and for ornamental purposes. When mature, the fragrant fruits are used to scent linens and clothes, are hung up as room deodorizers, and are said to repel insects. They are also made into preserves, but the immature fruits are better for cooking and eating raw. Various plant parts have medicinal applications.

Casabanana is propagated from seeds or cuttings, and the vines trained on sturdy trellises or other supports. It requires high temperature for good fruit ripening. A ripe fruit can be stored for several months.

Round Melon

Praecitrullus fistulosus, the only species in its genus, was formerly classified as a variety of watermelon. It differs from watermelon, however, in chromosome number and pollen characteristics, and will not cross with species of *Citrullus*.

The monoecious vines have branched tendrils. Male flowers are in clusters and female flowers are solitary. Immature fruits are hispid, whereas mature fruits are glabrous. The round, green fruit reaches 10 cm in diameter. Seeds are black with a ridged margin.

Round melon is a popular vegetable in northern India, where it is believed to have originated. In India and Pakistan, the fruits are harvested when fully grown, but before the seed coats harden. The fruit is cooked, often with lentils, after removing the seeds. Round melons are also pickled or prepared as preserves. The roasted seeds are edible as well.

5

CULTURAL REQUIREMENTS

INTRODUCTION

Specialized cultivation techniques for cucurbit crops are almost as old as the crops themselves. From a Chinese agricultural encyclopedia of the sixth century AD comes this description of planting melons, as translated by Shih (1962):

> How to plant melons: Wash the seed with water. Mix with table salt (treatment with table salt protects the vine from mildew). Slice off the topmost layer of dry soil with shovel. (If not sliced off first, no matter how big the hole is, there is always some dry soil in it, and the seeds will not sprout.) Then make a pit as big as a bowl. Place 4 melon seeds and 3 soya beans in the sunny side of the pit. After several leaves have expanded on the melon vine, pinch off the bean seedlings. (Melon seedlings are too weak to break through the ground, therefore advantage should be taken of the lifting force of sprouting beans. When melon vines have grown somewhat, the bean-stalks must be got rid of so that the vine may not be shadowed to disadvantage. – Bean-stalks when pinched will give out sap to moisten the soil. Never pull, or the ground would be loosened and dry.)

Even today, the diversity in cucurbit crops is reflected by a diversity of regional cultivation practices. Although many of these practices are converging towards efficacious commercial production techniques, differences among the crops, even at the level of cultivars, requires that individual attention be given to the cultivation of specific cucurbits. Consequently, cultural practices relating to the minor crops are primarily described with that crop in Chapter 4. In this chapter, we discuss research on the cultural requirements of cucurbits, most of which has been conducted on the major crops, and the commercial growing practices associated with these crops.

EFFECTIVE SEED GERMINATION

Seed treatment can help prevent damping off and some other diseases. Seed companies often coat cucurbit seeds with fungicide and adjuvants to enhance germination and vigour. Hot water (ca 55°C) seed treatment controls some seedborne diseases, as well as promoting germination in freshly harvested seeds.

Time to germination is usually longer and germination percentages lower for the minor cucurbit crops (except chayote) than for the major crops. When germinating cucurbit seeds indoors, they should be kept lightly moist, but not too wet, and incubated at low light levels or in darkness. Generally, seed germination is inhibited at temperatures below 15°C and is rapid at 25–30°C. Cultivars within a species can vary greatly, however, in response to temperature. Bushy birdsnest-type melons have relatively large seeds that germinate more quickly and in much higher percentages at 15°C than seeds of other cultivars (Fig. 5.1). Germination of the birdsnest

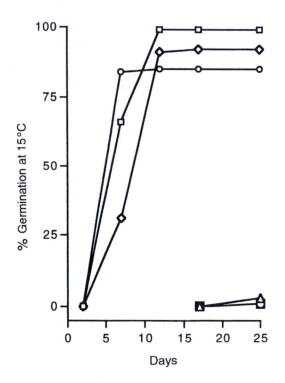

Fig. 5.1. Per cent germination at 15°C of three lines of birdsnest-type melons (top group of symbols) and three other melon cultivars ('Noy Yizre'el', 'Top Mark', and 'U.C. Perlita Bush'; bottom group). Two additional cultivars ('Perlita' and 'U.C. Top Mark Bush') tested did not germinate within 25 days. (Data from Nerson et al., 1982.)

lines can be enhanced further by several treatments, including presoaking the seeds for 24 h in aerated distilled water or in aqueous solutions of 3% KNO_3, 0.1 mmol l^{-1} 6-benzylamino purine, or 1 mmol l^{-1} gibberellic acid (Nerson *et al.*, 1982).

Cucumber seeds imbibed for 6 days at 2.5°C were still able to germinate normally in recent experiments, although emerging radicles were irreversibly damaged after 96 hours at this temperature (Jennings and Saltveit, 1994). At 25°C, imbibition appeared complete between 12 and 16 h. Longer imbibition periods at this temperature produced seedlings that were more sensitive to subsequent chilling at 2.5°C (Fig. 5.2). The mean germination rate (i.e. the inverse of mean time to germinate) for seeds of 'Dasher II' and 'Poinsett 76' at temperatures ranging from 5 to 30°C was similar for the two cultivars, but significantly higher than for seeds of 'Poinsett 76' that were 3 years older (Fig. 5.3).

In another experiment (Benzioni *et al.*, 1993), seeds of African horned cucumber germinated best at 20–35°C and were unaffected by salinities of up to 50 mmol l^{-1} NaCl. Germination was completely inhibited at 8°C. Although there was eventually 90% germination at 12°C, maximum germination was

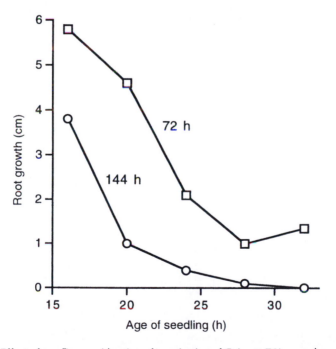

Fig. 5.2. Effect of seedling age (duration of germination of 'Poinsett 76' cucumber seeds at 25°C) on root growth after germinated seeds were chilled at 2.5°C for 72 or 144 h. (Redrawn from Jennings and Saltveit, 1994, with permission from the American Society for Horticultural Science, 600 Cameron Street, Alexandria, VA 22314, USA.)

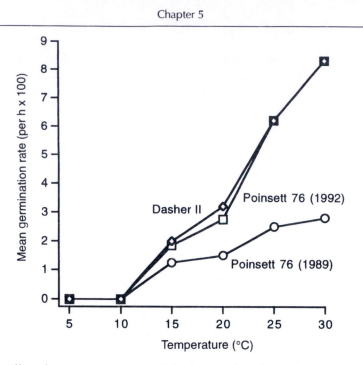

Fig. 5.3. Effect of temperature on mean germination rate of 'Dasher II' and 'Poinsett 76' cucumber seeds from 1992 and 'Poinsett 76' seeds from 1989. (Redrawn from Jennings and Saltveit, 1994, with permission from the American Society for Horticultural Science, 600 Cameron Street, Alexandria, VA 22314, USA.)

not reached until the 24th day of treatment versus the third day when seeds were kept at 25°C.

Freshly harvested cucurbit seeds may be dormant. Dormancy-breaking treatments, when needed, include scarifying, cracking or removing the seed coat. Clipping the radicle end of the seed can be promotory in bottle gourd and some species of squash. Pre-treatment with gibberellins may aid germination in cucumber, melon, bitter melon and other cucurbits. Pre-drying is helpful for bottle gourd, cucumber and the squash *Cucurbita pepo*, whereas pre-soaking is advised for angled loofah and watermelon. For recalcitrant seeds of *Cucumis*, pre-chilling at 5°C for several weeks is suggested. In cucumber, secondary dormancy may be induced when trying to germinate seeds under unfavourable conditions.

LAND PREPARATION

A previous field crop is usually ploughed under after harvest and given time to decay (over winter in temperate areas). Squash and cucumber have been grown successfully by no-till planting, in which the previous crop and weeds are not ploughed under, but instead, are killed with a herbicide and the cucurbit seeds planted through the stubble (Pierce, 1987).

Tractors and other heavy equipment should not be used on wet fields for soil preparation or other operations. Rototillers of light weight are commonly used for soil preparation and cultivation in home gardens everywhere as well as on small, commercial farms in Asian countries.

A winter cover crop can be planted in the autumn to add organic matter to the soil and help prevent erosion. About a month before planting the new cucurbit crop, the field should be deeply ploughed and allowed to settle. A fine-grained, flat surface is achieved by mechanical disking and harrowing. The field should be relatively free of stones. Levelling may be needed for furrow irrigation.

Raised beds (10–50 cm high) improve drainage, modify temperature and increase the depth of the rooting zone. They can be laid out with a double moldboard plough to which a marker is attached. The plough makes 25-cm deep furrows and the bed is created by back-furrowing with a turn-plough. The south side of the bed is disked and harrowed and smoothed with a V-drag.

Before planting, the ploughed field can be treated with herbicide to kill germinating weeds, particularly perennial grasses. After the weeds have died, the soil may be fertilized with organic material or inorganic preparations. Inorganic fertilizers are broadcast before beds are formed, banded into the beds between the seed-row and the irrigation furrow, or added to the irrigation water. Irrigation should be applied, if needed, soon after planting and should soak the top 1–2 m of soil.

PLANTING, THINNING AND SPACING FOR OUTDOOR CROPS

Cucurbits are generally planted in the field after the danger of a late frost has passed. Winter squash, pumpkin, melon, watermelon, bottle gourds and loofah require long growing seasons (Table 5.1). Cucumber, summer squash and bitter melon grow quickly in warm weather, requiring less than 2 months from seeding to the harvest of edible immature fruits. In California and other areas with a long growing season, it is possible to double crop and have two plantings of mechanically harvested pickling cucumbers in the same field. In the frost-free tropics, two crops per year can be produced for wax gourd and mature chayote plants will bear fruit year-round.

Monoculture of cucurbits is prevalent today. In the past, Native Americans interplanted squash with corn, and this is still practised in parts of Latin America. Intercropping is also practised in Africa, where fluted pumpkin is grown with cassava and yams, and egusi is interplanted with grain crops and yams.

Table 5.1. Seed size, fruit maturation time, and recommended spacing for major cucurbit crops.

Crop	Number of seeds per gram	Days from planting to harvest	Days from pollination to market maturity	Spacing (cm)	
				Within row	Between rows
Cucumber, pickling	390	48–58	4–5	20–30	90–180
Cucumber, slicing	390	62–72	15–18	20–30	90–180
Melon	460	85–110	42–46	30	150–210
Pumpkin	4–11	100–120	65–110	90–150	180–240
Squash, summer	5–14	40–50	3–6	60–120	90–150
Squash, winter	5–14	85–110	55–90	90–240	180–240
Watermelon	7–11	75–95	42–45	60–90	180–240

Modified from *Knott's Handbook for Vegetable Growers*, © 1980, O.A. Lorenz and D.N. Maynard. Reprinted by permission of John Wiley & Sons, Inc.

Direct seeding in the field is customary. In less mechanized societies, planting is often by the hill system, whereby three to ten seeds are sown by hand in the same place and later thinned to one to three plants per hill. Although home gardeners and others still plant using this method, most large-scale growers in developed countries now sow cucurbits with tractor-mounted precision seeders, which uniformly space the seeds, eliminating the need for thinning.

In short-season areas, the season can be extended by sowing the seeds indoors and transplanting young plants to the field. Transplants should be only about 3 weeks old, and they should not be bare rooted; instead, they should be grown in peat pots or other containers that permit transplanting without disturbing the roots. The use of transplants early in the season, especially when plastic mulch and rowcovers are available, can hasten maturity and help the grower to market his crop early in the season when prices are usually higher.

The rate of seedling emergence is influenced by the depth at which seeds are sown. Growers planting very early in the season sometimes sow cucurbit seeds at two different depths. The shallow planting comes up first and can be harvested for the early market unless the seedlings are killed by a late frost, in which case they can be replaced by the deeper planting emerging after the frost.

Summer squash cultivars generally have a bush habit and are grown at much closer spacings than vining winter squash and pumpkin plants (Table 5.1). Bush cultivars with short internodes and a compact plant habit are also available for melon, cucumber and watermelon; they are grown by home gardeners with limited space, but are seldom raised on commercial farms. Small bush cucumbers (e.g. 'Baby Bush') can be placed as close together as every 5 cm within rows. Suggested within-row and between-row spacings of vining plants of the major cucurbit crops are listed in Table 5.1. Field-grown bottle gourd plants should be sown 1–3 m apart.

Increasing plant densities has been used to enhance cucurbit fruit yields.

Table 5.2. Effect of plant spacing on fruit number and leaf area for four genotypes of melon.

Parameter and genotype	Plant spacing (cm)	
	60	30
Number of fruits per plant[1]		
'Mainstream'	5.8	3.6
Main Dwarf	5.0	3.8
Ky-P$_7$	1.8	1.4
'Bush Star'	1.4	1.5
Significance		
Spacing	***	
Genotype	***	
Interaction	*	
Number of fruits ('000) per hectare		
'Mainstream'	35.1	89.6
Main Dwarf	36.4	90.8
Ky-P$_7$	13.8	35.5
'Bush Star'	8.9	30.7
Significance		
Spacing	***	
Genotype	***	
Interaction	ns	
Leaf area ('000 cm^2) per plant[2]		
'Mainstream'	15.5	10.5
Main Dwarf	10.8	7.0
Ky-P$_7$	17.0	11.1
'Bush Star'	11.8	5.4
Significance		
Spacing	**	
Genotype	**	
Interaction	ns	

[1]Means for five harvests.
[2]Calculated at 72 days after seeding.
ns, *, **, *** = Non-significant or significant at $P \le 0.05$, 0.01, or 0.001, respectively.
Modified from Knavel (1991), with permission from the American Society for Horticultural Science, 600 Cameron Street, Alexandria, VA 22314, USA.

With melon, high density planting generally produces a greater number of fruits per area, whereas low density planting can lead to a larger percentage of unmarketable sunburned fruits. However, as plant density increases, fruit quality (i.e. soluble solids content) and fruit number per plant typically decrease, as a result of reduced leaf area per fruit. Comparing within-row spacings of 30 cm and 60 cm between melon plants, Knavel (1991) found that higher density was significantly correlated with lower total leaf area per plant, lower total plant dry weight, fewer fruits per plant (although fruit size was unaffected) and a larger number of fruits per hectare (Table 5.2). Genotype effects reflected differences among the cultivars tested with respect to canopy architecture: longer internodes and smaller leaves in 'Mainstream' and genotype Main Dwarf provided a greater percentage of plant leaf area exposed to sunlight, whereas less secondary stem branching in genotype Ky-P$_7$ meant fewer potential fruiting sites on the plant.

Although plants of the bushy 'Autumn Pride' (*Cucurbita maxima*) have been grown as close together as 30×150 cm (22,000 plants per hectare), plants at this high density produced squashes too small to be marketable and some were misshapen (Loy and Broderick, 1990). In contrast, varying population density from 50,000 to 850,000 plants per hectare did not affect the percentage of misshapen cucumbers for the pickling cultivars 'Bounty' and 'Premier' (Cantliffe and Phatak, 1975). For these cultivars, significant increases in fruit production per area were achieved when planting density was increased from 250,000 to 500,000 plants per hectare (Table 5.3). As

Table 5.3. Effect of plant density on fruit yield for two cultivars of field-grown pickling cucumbers.

Plants per hectare	Spacing (cm)	Number of fruits per plant		Tons of fruit per hectare[1]	
		'Bounty'	'Premier'	'Bounty'	'Premier'
850,000	10×10	1.7a[2]	1.3a	32.9b	52.3d
650,000	13×13	2.0a	1.8a	30.7b	52.3d
500,000	15×15	2.1a	2.0a	32.9b	43.3cd
250,000	19×19	2.9ab	2.8b	24.0ab	34.4bc
200,000	23×23	2.9ab	3.3b	20.3a	33.2b
150,000	25×25	4.7b	5.1c	23.5ab	36.3bc
100,000	31×31	5.5b	5.5cd	17.9a	31.9b
50,000	46×46	8.5c	6.3d	19.1a	23.0a

[1]Harvest made 49 days after seeding.
[2]Mean separation within columns by Duncan's multiple range test at the 5% level.
Modified from Cantliffe and Phatak (1975), with permission from the American Society for Horticultural Science, 600 Cameron Street, Alexandria, VA 22314, USA.

with melon, the number of fruits per plant decreased with increasing population density.

In another pickling cucumber experiment, this one involving an indeterminate vine cultivar ('Tamor') and a determinate, highly branched bush cultivar ('Castlepik'), Widders and Price (1989) found a high correlation between leaf lamina dry weight and fruit growth rate (Fig. 5.4), supporting the hypothesis that net photosynthetic capacity limits a plant's fruit production potential. Leaf area and fruit number and weight per plant decreased significantly when planting densities were increased from 44,000 to 194,000 plants per hectare (Fig. 5.5). Within-row spacing effects (29, 14 and 11 cm between plants) were more significant than between-row spacing effects (71 and 36 cm). Unlike the results of Knavel's (1991) study on melon and Loy and Broderick's (1990) study on squash, genotypic differences in plant architecture did not significantly affect fruit production per cucumber plant.

In high density situations, limited and insufficient resources (e.g. water and nutrients) are expected to create competition among neighbouring plants, tending to decrease an individual plant's production. At later stages of

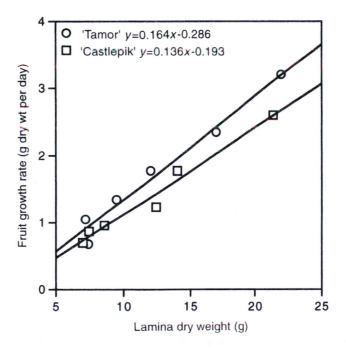

Fig. 5.4. Relationship between total leaf lamina dry weight and net fruit growth rate per plant at 48 days after planting for two cultivars of pickling cucumbers. Correlation coefficients for 'Castlepik' and 'Tamor' are $r^2=0.985$ and $r^2=0.991$, respectively. (Redrawn from Widders and Price, 1989, with permission from the American Society for Horticultural Science, 600 Cameron Street, Alexandria, VA 22314, USA.)

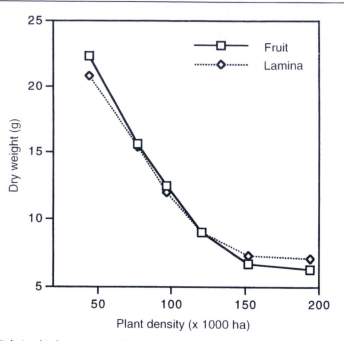

Fig. 5.5. Relationship between population density and the dry weights of fruit and lamina tissues of individual pickling cucumber plants as determined at 48 days after planting. Weight values are means of two cultivars × two replications, and were calculated at six different plant densities. (Modified from Widders and Price, 1989, with permission from the American Society for Horticultural Science, 600 Cameron Street, Alexandria, VA 22314, USA.)

maturity, especially during fruiting, canopy overlap and shading create competition for light as well. Given that unit leaf rate (total above-ground plant dry weight produced per day divided by the total lamina dry weight) was unaffected by spacing in their experiments, Widders and Price (1989) proposed that foliar competition did not alter photosynthetic efficiency of the leaves, but instead, influenced carbon partitioning among alternative sinks within the plant.

Competition, allelopathy, and various environmental and genetic factors influence the success of using higher plant densities to enhance overall crop yield. Reduction in individual plant yield and cost of seed must be weighed against the increase in plant number per area. For once-over harvested pickling cucumbers, additional concerns include endogenous and exogenous factors that affect the sink strength of early fruits and the number of fruits that can be set simultaneously.

Rows of vertical trellises for loofah production should be at least 2 m apart to accommodate equipment. The results of Davis (1994) on the effects of within-row trellis spacing and pruning on loofah sponge size and quality are

Table 5.4. Effect of within-row spacing and pruning on size and quality of loofah sponges.

Main effect	Mean sponge size		Number of sponges per hectare (1000s)				Sponge quality attributes		
			Diameter (cm)		Length (cm)				
	Diameter (cm)	Length (cm)	5.1–7.6	7.7–10.2	15.2–30.5	30.6–45.7	Fibre density[1]	Sponge strength[2]	Sponge appearance[3]
Within-row spacing (cm)									
30.5	7.9	39.9	36.2	31.2	10.7	42.6	2.4	2.2	2.6
61	8.3	42.2	24.0	32.4	7.4	31.7	2.6	2.3	2.4
91	8.6	42.7	15.7	33.7	4.1	28.8	2.9	2.6	2.9
LSD$_{0.05}$	0.02	1.5	5.9	ns	3.0	5.8	0.4	ns	0.4
Pruning									
Topping at node 6	8.3	42.7	22.1	30.9	5.5	31.0	3.0	3.0	2.9
Removing four laterals	8.2	40.6	28.6	31.1	8.5	37.2	2.6	2.0	2.6
No pruning	8.3	41.6	25.2	35.3	8.1	34.8	2.3	2.1	2.4
LSD$_{0.05}$	ns	1.5	ns	ns	ns	ns	0.4	0.4	0.4
Significance									
Within-row spacing	**	**	**	ns	**	**	*	ns	*
Pruning	ns	*	ns	ns	ns	ns	**	**	*
Interaction	ns	ns	ns	ns	ns	ns	ns	ns	ns

[1] Overall fibre density rated on a scale of 1 to 5 (1 = very loose, open weave of fibres through which newspaper print could be read; 5 = very closely woven fibres that allowed little light to penetrate).
[2] Sponge strength was rated on a scale of 1 to 5 (1 = crumbled easily; 2 = did not crumble but could be pulled apart; 3 = took some effort to pull apart; 4 = was very difficult to pull apart; 5 = could not be pulled apart).
[3] Visual appeal was rated on a scale of 1 to 5 (1 = poor with loose, broken, uneven fibres; 3 = looked good with fairly closely woven, even fibres; 5 = excellent with very closely woven, even fibres and a smooth finish).
LSD = least significant difference.
ns, *, ** = Non-significant or significant at $P \leq 0.05$ or 0.01, respectively.
Modified from Davis (1994), with permission from the American Society for Horticultural Science, 600 Cameron Street, Alexandria, VA 22314, USA.

presented in Table 5.4. Sponge size, fibre density and sponge appearance scores were greater for plants spaced 91 cm apart than for spacings of 30.5 or 61 cm. However, the highest marketable yields were obtained with the 30.5 cm spacing regime.

Chayote plants should be spaced 3–4 m apart on trellises. Bottle gourd, snake gourd, fluted pumpkin and wax gourd plants can be placed every 60–80 cm along the trellis, and bitter melon plants can be placed as close as 30 cm apart.

MULCHES AND PLANT COVERS

Plastic mulch is used to warm the soil, conserve moisture, reduce leaching of nutrients, and provide protection against soil pathogens and fruit rot. Clear plastic increases soil temperature more than black plastic, but black polyethylene plastic is used more often in order to control weeds. Infrared-transmitting plastic permits infrared rays to penetrate the soil, thereby increasing soil temperature intermediate to that with black or clear plastic mulch, and yet still preventing weed growth by the lack of other wavelengths of light.

Aluminium mulch or aluminium bonded to paper reflects heat as well as light, and does not warm the soil as well as black or clear plastic. However, aluminium and other brightly coloured mulches have the advantage of repelling aphids and some other insects, thereby reducing losses from viral diseases transmitted by these insects.

Plastic mulch can be applied before planting with a tractor-mounted implement. The plastic covers the row for a width of 1 m or more; at intervals, small openings are made in the plastic to permit insertion of seeds or transplants into the soil. The openings are generally sufficient for water absorption from rain or irrigation, but it is recommended that the plastic be applied only when there is adequate soil moisture. Plastic mulch is often used in combination with rowcovers and drip irrigation.

Polyethylene plastic needs to be picked up and disposed of at the end of the season. Biodegradable and photodegradable plastic mulches have been developed to avoid this problem.

In the USA, hot tents (small paper coverings over individual plants), often supplemented with windbreaks of brush, have been used to protect melon plants early in the season, when it may be cool and windy. Similarly, strips of rye, sorghum or other cereal can be sown in rows at right angles to the prevailing wind to protect cucurbit seedlings.

For additional protection from outdoor elements, especially cold, a variety of enclosures exist which extend the effective growing season of cucurbits in temperate areas. For example, home gardeners may start their melon plants in small framed glass structures which accommodate only a few mature plants. The soil or other medium in the frame is warmed by sunlight and the absorbed heat keeps the plants warm during cold spring nights. Glass covers over plant rows, called cloches, have been used successfully in the UK for many years to increase local temperature and ward off frost.

More recently, unheated plastic tunnels have become popular in Mediterranean countries and elsewhere for commercial production of melon, cucumber, summer squash and other cucurbits. Polyethylene or polypropylene plastic is typically suspended by wire hoops over a row of plants to form these rowcovers. Floating rowcovers require no support: lightweight fabric rests on the plants, with the edges secured by soil to keep it in place. Tunnels and floating rowcovers increase temperature during the day, protect young plants from the wind, and exclude harmful insects and insect-transmitted diseases. However these coverings provide only slight protection against frost.

The temperature may build up in plastic tunnels so much on warm, sunny days that cucurbits are harmed unless ventilation is provided. Some growers attach strips of plastic to wire hoops with clothes pins or other devices so that the plastic can be opened on warm days. Others use plastic with slits for ventilation. The rowcovering should be removed or opened before cucurbits flower to permit insect pollination, unless a parthenocarpic cultivar is grown.

Soltani *et al.* (1995) evaluated various mulch and rowcover combinations for their effects on growing watermelon. Although light intensity was lower under rowcovers, the three materials used (clear polyethylene, spunbonded polypropylene polyamide net, and spunbonded polyester fabric) transmitted enough of photosynthetic photon flux (70–80%) that the light saturation point of the plants was probably exceeded. Although CO_2 concentration was higher inside the mulch transplanting holes (600 µl l^{-1}), foliage level concentrations were normal, apparently due to rapid air flow inside the tunnels. Consequently, temperature was deemed the most important environmental factor modified by the treatments. Compared with bare ground, soil temperatures were significantly higher under mulches (ca 2–4°C higher at 10 cm soil depth under black mulch) and more so (by an additional 1–3°C) when rowcovers were used with the black mulch. As a result of warmer temperature, plants under rowcovers (especially those made of polyethylene or polyester) grew faster, flowered sooner and yielded more fruits.

GLASSHOUSE CULTURE

The largest protective structures used for commercial cucurbit production are glasshouses, which broadly speaking, include angled-roof buildings constructed of aluminium and glass, structures made of corrugated fibreglass and rigid or air-inflated houses enclosed with polyethylene. Glasshouses typically possess equipment to control one or more of the following environmental conditions: supplementary lighting (e.g. high pressure mercury lamps), shading (e.g. inner or outer blinds), heating (e.g. hot water boilers), cooling (e.g. evaporative cooling pads), ventilation and humidity (e.g. mechanically controlled ceiling vents), and CO_2 concentration. Irrigation setups range from overhead sprinklers to drip fertigation systems to sophisticated hydroponics.

Glasshouse cultivation is suitable for most cucurbits when trellising or bush cultivars can be used. In Japan and Korea, some melon and watermelon cultivars are commonly raised in glasshouses. However, the most suitable cucurbit crop for growing in a controlled environment is cucumber (Fig. 5.6). In fact, cucumber may have been the first crop protected from unfavourable outdoor weather; the plants were grown in forced culture by the ancient Romans. Today, glasshouse cucumber is an important crop in The Netherlands, the UK, China, Japan, Korea and the Middle East. Most of the discussion of glasshouse culture that follows is based on experience with and research on cucumber.

To conserve glasshouse space and heating, cucumber is often grown from transplants. Seedlings are cultivated in small containers to about 10 cm high, and then transplanted at a wider spacing (ca 40 cm apart within a row) for

Fig. 5.6. Young cucumber plants growing in soil-filled troughs in a glasshouse in China. When the plants are a little older, the plastic strips hanging from above will be positioned so as to train the vines upward.

Fig. 5.7. Cucumber plants growing on straw bales at the Glasshouse Crops Research Institute, UK.

trellising (see 'Support, training and pruning', this chapter, page 142). Given their climbing nature, these plants are usually raised to maturity in beds, boxes, troughs, pots or bags which sit on the clean glasshouse floor or on low slatted benches. The roots are anchored in sterilized soil or a soilless medium of peat, vermiculite, perlite, straw, sawdust, sand or other substrate, such as rockwool, a relatively inert material composed of coke and limestone made molten at 1600°C, spun into fibres, and then woven into slabs.

The use of rockwool culture is popular in The Netherlands and the UK. The slabs are laid on grooved polystyrene boards in which a plastic water pipe is set for root zone warming. For planting, holes are cut on top of the slabs through the polyethylene sleeve. Frequent irrigation is necessary with rockwool, and root zone pH should be kept at 5.5–5.7.

In straw bale culture (Fig. 5.7), fertilizer is added to water-soaked bales, which are then subjected to at least 15°C to promote fermentation. After bale temperature peaks near 40°C and starts to fall, cucumber plants are rooted in a soil or peat mixture placed on top of the bales. Straw from wheat or barley crops treated with herbicides should be avoided as the residues can harm cucumber plants.

With some growing systems (e.g. peat modules and rockwool slabs), nutrients can be supplied with the irrigation water via a fertilizer injector in an 'open' watering system. In a closed hydroponic setup, the aqueous nutrient solution is recycled, monitored and adjusted as necessary. For example, Japanese producers raise cucumber and melon seedlings in perlite for 4–6 weeks until the young plants can support themselves; then, the plants are inserted into the styrofoam lids of solution-filled troughs (Resh, 1987).

The nutrient film technique (NFT) is a hydroponic setup which strives to improve aeration by constantly circulating a shallow nutrient solution across the cucumber root zone. The solution is pumped from a catchment tank up to the raised ends of sloped troughs made of polyethylene film; the solution flows through the troughs past the roots of the inserted plants and back into the tank. NFT systems can be complex, with solution heating tubes to maintain a high root temperature (at least 18°C and preferably 25–29°C for cucumber), CO_2 enrichment lines, collecting troughs, and electronic control devices for monitoring the pH (5.5 is good for cucumber) and electrical conductivity of the solution.

Glasshouse temperatures should be no lower than 16°C at night, ranging up to 30°C during the day, depending on the crop and stage of growth. An average daytime temperature of 21–22°C, with 2°C less at night and venting at 27°C, is standard glasshouse practice for winter cucumber crops in northern Europe. After plants are well established, mean temperature can be dropped 2–3°C. Heating to higher temperatures, which would produce greater yields, is not considered cost effective when outdoor temperature is low (Liebig, 1980). However, higher growing temperatures are possible for summer crops and are recommended for glasshouse-grown melon.

Ventilation is required, especially in summer, to reduce glasshouse temperature and humidity. High humidity, which builds up as a result of transpiration, promotes fungal growth. At the end of a hot day, venting prevents an increase in relative humidity during a subsequent cold night. If glasshouse temperature rises too quickly in the morning, venting is needed to avoid condensation on the cool fruits. Air movement that encourages transpiration is desired and accomplished with passive ventilation through roof openings or with the aid of fans.

Supplementary lighting and carbon dioxide enrichment in the glasshouse help to increase cucumber yield and reduce time to harvest. Fluorescent or mercury-fluorescent lamps are lit over cucumber seedlings for 12 or more hours per day for the first 3 weeks of growth, producing larger and hardier seedlings. Supplementary lighting can be continued until the flowering stage, at which time adjustments in daily light duration may be needed to accommodate the plant's sensitivity to daylength (see 'Daylength and light intensity,' this chapter, page 134).

The benefits of supplementary lighting are enhanced further by the concurrent application of supplemental CO_2, since both are necessary resources for photosynthesis. For the pickling cucumber 'Wisconsin SMR 18', the greatest yield in dry weight of young plants was obtained at the highest levels tested: 2150 ppm CO_2 combined with light at 6000 μW cm^{-2} (Hopen and Ries, 1962). As expected, the interaction of light and CO_2 was significant. Plants grown to fruiting developed a greater number of marketable fruits at higher light levels.

The effect of CO_2 enrichment alone was evaluated for glasshouse cucumber crops grown in North Carolina (Peet and Willits, 1987). Both extended CO_2 duration (by the use of rockstorage instead of venting to control temperature) and greater CO_2 intensity were tested. At short durations (<6 h day^{-1}), concentrations of 5000 μl l^{-1} produced up to a 30% increase over controls (i.e. no CO_2 enrichment) in marketable fruit yields of spring crops. At longer enrichment durations, maximum results were reached at lower concentrations (600–800 μl l^{-1}).

In commercial cucumber production, standard practice is to increase glasshouse CO_2 levels to 1000 μl l^{-1} during those daylight hours when vents are closed or nearly so; higher levels are not considered cost effective. Carbon dioxide gas can be obtained and injected directly into the glasshouse, or produced by burning low-sulphur kerosene, paraffin, propane or natural gas. Cucumber plants grown on bales of decomposing straw, which emit CO_2 at nearly 1000 μl l^{-1}, do not need supplemental CO_2 (Hand, 1984).

A distinction of most glasshouse cucumbers is their ability to set early parthenocarpic fruits without pollination. A single incompletely dominant allele, plus modifying genes, is the basis for this parthenocarpy. These alleles were derived from European germplasm developed more than a century ago; breeders selected high yielding types in their winter glasshouses, without realizing that genetic parthenocarpy was the reason for the high yields produced by plants grown in the absence of pollinating insects. Also, parthenocarpic fruits are less inhibiting to additional fruit formation on a plant than are fruits with seeds.

Cucumber yields in glasshouses can be considerably larger than those in the field. Contributing factors are parthenocarpy, closer spacing, more favourable climate and longer harvest period. Three to five glasshouse crops can be produced a year. Or, one long-duration crop can be grown in the glasshouse from February through October by using the renewal umbrella system of trellising (see 'Support, training and pruning', this chapter, page 142).

GRAFTING

Cucurbit grafting has a 50-year history in Japan and Korea. Most watermelon, melon and glasshouse cucumber crops are grafted in these countries (Table 5.5). Grafting is also practiced in other parts of Asia (e.g. China; Fig. 5.8) and Europe where land use is very intensive, but seldom in the USA.

Grafting a crop to the rootstock of another cucurbit cultivar or species can

Table 5.5. Annual statistics for the total area cultivated in cucurbits and the area cultivated with grafted seedlings for Japan and Korea in the early 1990s.

	Japan		Korea	
Cucurbit crop	Total cultivation area ('000 ha)	Area of grafted plants ('000 ha)	Total cultivation area ('000 ha)	Area of grafted plants ('000 ha)
Watermelon				
Field	24.9	23.8	28.3	26.9
Glasshouse	3.2	3.2	7.4	7.4
Cucumber				
Field	14.8	4.6	3.8	0.4
Glasshouse	7.0	6.0	4.7	3.3
Melon				
Field	5.5	3.7	3.9	3.3
Glasshouse	3.4	2.5	5.2	5.0

Modified from Lee (1994), with permission from the American Society for Horticultural Science, 600 Cameron Street, Alexandria, VA 22314, USA.

reduce susceptibility to soilborne diseases, increase low temperature tolerance and enhance water and nutrient uptake, which promotes growth and extends harvest time (Table 5.6). Fruits of grafted watermelon can be significantly larger than those of intact plants. Rootstocks must be carefully chosen, however, because they may have additional undesirable effects on the grafted scion. For example, a cultivar-specific trait such as fruit colour in cucumber can be modified by the biochemical products of the rootstock.

Cucurbits are usually grafted before expansion of the first leaves in rootstock and scion seedlings. The following grafting descriptions are from Lee (1994). 'Hole insertion' grafting is frequently used for watermelon scions because of their small size as seedlings. The apical bud is clipped out of the rootstock; the watermelon seedling is severed just above the roots and then inserted into the hole between the cotyledons of the rootstock. The 'tongue approach' is often used with cucumber seedlings because of their hearty hypocotyls. A downward and upward cut is made in the hypocotyl of the rootstock and scion, respectively. The 'tongue' of the scion is then inserted into the tongue of the rootstock, and a grafting clip secures the area. After about a week, the scion's hypocotyl is cut below the graft union. 'Cleft grafting' is employed for older seedlings with leaves. A cleft-shaped segment is removed from the side of the top of the severed rootstock stem; the scion is clipped above the root area, placed in the cleft of the rootstock, and secured with a special grafting clip. All grafted seedlings need to be monitored for

Fig. 5.8. Cucumber seedling grafted to Malabar gourd rootstock in Chinese glasshouse. Note grafting clip holding the hypocotyls together below the large cotyledons of the rootstock.

7–10 days after grafting. As commercial seedling production continues to expand, automated methods of grafting and post-grafting care are being developed.

TEMPERATURE

Cucurbits are very sensitive to low root-zone temperature, which increases susceptibility to diseases. At temperatures below 20°C, water uptake may be restricted and the plant injured or killed by drought, even though soil moisture is ample. When sowing cucurbits on ridges or raised beds, planting on the southern slope of the bed will give the seedlings the advantage of somewhat warmer soil temperatures. For winter glasshouse crops, water absorption may be restricted by the low temperature of the irrigation water. To prevent this problem, some growers heat the cold irrigation water or blend

Table 5.6. Cucurbit rootstocks, grafting methods and purposes of grafting for cucurbit crops.

Cucurbit crop	Popular rootstocks	Grafting methods[1]	Purposes[2]
Watermelon	Bottle gourd	1	1,2
	Interspecific hybrids[3]	1,2	1,2,3
	Wax gourd	1,3	1,2
	Cucurbita pepo	2,3	1,2,3
	Cucurbita moschata	1,2	1,2,3
	Sicyos angulatus	2	5
Cucumber	Malabar gourd	2	1,2,3
	Interspecific hybrids[3]	1,2	1,2,3
	(*Cucurbita maxima* × *C. moschata*) F_1	2	1,2,4
	Cucumber	2	1,2
	Sicyos angulatus	2	2,5
Melon	Interspecific hybrids[3]	2	1,2,3
	Cucurbita moschata	2	1,2,3
	Melon	2,3	1,3,4

[1]Grafting methods: 1 = hole insertion; 2 = tongue approach; 3 = cleft grafting.
[2]Purposes of grafting: 1 = fusarium wilt control; 2 = growth promotion; 3 = low temperature resistance; 4 = growth period extension; 5 = nematode resistance.
[3]Many interspecific hybrids are commonly obtained by fertilized ovule culture *in vitro.*
Modified from Lee (1994), with permission from the American Society for Horticultural Science, 600 Cameron Street, Alexandria, VA 22314, USA.

it with warm water. Increasing the temperature of the medium around the roots is also practised. This can be accomplished with a circulating hot water system constructed of polyethylene pipes buried in the medium. Water at 35–40°C is used to maintain root medium temperature at 20–23°C.

Kramer (1942) investigated the relationship between soil temperature and transpiration in watermelon plants, finding that only 20 and 50% as much water was absorbed at 10 and 15°C, respectively, as at 25°C. Decreased absorption at low temperatures appears to be the result of several factors, including decreased root growth, lower respiration rates, increased viscosity of colder water and decreased permeability of root cell membranes. Root permeability in watermelon decreases when temperatures fall below 22°C, with water uptake decreasing most rapidly between 18 and 16°C.

Cucurbit plants are subject to chilling injury when exposed to temperatures less than 10°C, causing yellowing of foliage and short, misshapen fruits. Cucumber fruits become pitted with small sunken spots and rot prematurely. Chilled winter squashes and pumpkins do not store well, but will rot prematurely also. Plants can be 'hardened', principally by adequate exposure to low but non-injurious temperatures, to somewhat reduce the

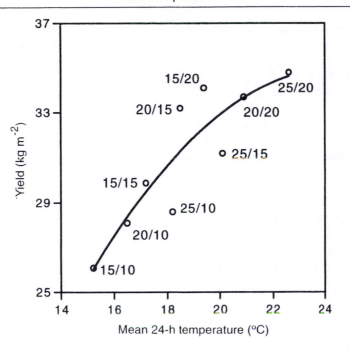

Fig. 5.9. Effect of mean 24-h temperature from various day/night temperature regimes on glasshouse cucumber yield after 20 weeks of harvesting. The numbers by each data point indicate the nominal day/night temperature setting for that treatment. (Data from Slack and Hand, 1983.)

temperature at which chilling injury occurs. For example, several days exposure to 12°C can harden cucumber plants to subsequent temperatures below 10°C.

Cucurbits require warm temperatures (20–30°C) for best growth, and the annual species are often the first vegetables to be killed by a frost. The annual stems of perennial species (e.g. pointed gourd) may be killed by frost also, but new vines sprout from surviving underground stem tissue when warm weather returns. Malabar gourd and stuffing cucumber are more cold tolerant than most cultivated cucurbits, being able to grow at temperatures only slightly above freezing. Among the squashes, *Cucurbita maxima* and *C. pepo* are more tolerant of low temperature than are *C. argyrosperma* and *C. moschata*, which thrive at high temperature. In fact, fruits of some cultivars of *C. maxima* and *C. pepo* taste sweeter when grown in temperate areas with cool nights. In the tropics, these species can be cultivated at elevations of up to 2000 m. Cucumber prefers warm conditions (see Glasshouse culture, this chapter, page 125), but can be grown at cooler temperatures than melon or watermelon. Melons are of best quality when raised at 25–30°C. Stable day/night temperatures promote rapid growth in watermelon.

In various experiments (e.g. Krug and Liebig, 1980), raising the 24-h mean temperature from 15 to 30°C caused cucumber stem and leaf growth rates to accelerate, time to first harvest to shorten, and an increase in total fruit yield (e.g. Fig. 5.9). Although the amplitude around the mean generally had little effect, young plants (<34 days old) responded better to higher day temperatures among regimes with the same 24-h mean (Slack and Hand, 1983). However, large, quick fluctuations should be avoided; a sharp drop in temperature can cause cucumber fruits to become thin around their middle, and a short period of low temperature can produce scars on developing fruits.

Grimstad and Frimanslund (1993) found that stem elongation in young cucumber plants was less when night temperature exceeded day temperature. As long as fruit yield is not affected, maintaining a relatively short height is desired for greater ease in transporting and transplanting young plants in the glasshouse.

Optimum temperatures contribute to crop yield primarily when they occur during the early part of the growing season. The developmental response to temperature decreases as the crop matures, with other factors, especially fruit load, affecting fruit growth and yield on older plants.

DAYLENGTH AND LIGHT INTENSITY

Growing cucurbits in extreme latitudes is complicated by the more variable photoperiods as well as the cool temperatures in these areas. Long days favour male over female flower production in many cucumber, wax gourd and squash cultivars. In other species (e.g. Malabar gourd and chayote), long days inhibit flower production altogether. Genetic breeding has created various daylength-insensitive cultivars within the major cucurbit crops, and many of the minor cucurbit crops are naturally day-neutral.

Temperate region growers wanting to produce fruits from short-day plants like chayote during the summer need to control the photoperiod of young plants. From about the third to the fourth month of growth, vines should be shaded in the morning and evening with a dark cloth over a sturdy frame to give the plants a daylength of 8 h. Vines can be allowed to grow under natural daylengths after flowering has begun (Aung et al., 1990).

High light intensity is needed for optimum yield of cucurbits. Although chayote will flower under glasshouse conditions in the winter, fruit set is poor because of the low light intensities. Furthermore, vigorous vine growth under these conditions can cause readsorption of the contents of developing fruits by the vine (Aung et al., 1990). Supplementary illumination in the glasshouse (see Glasshouse culture, this chapter, page 125) is beneficial in areas and seasons with much cloudy weather. In the field, aluminium or other reflective mulch on the soil reflects additional light to the leaves of cucurbit crops.

Sunburn injury of immature fruits happens when cucurbit vines are moved aside during harvesting, exposing previously shielded, tender fruits to high light intensity. A white or tan necrosis develops on the part of the fruit suddenly exposed to sun.

SOIL

Cucurbits benefit from well-drained soils with a high organic content. Sandy or sandy loam soil is preferred for early production, but cucurbits can be grown on a variety of soil types. Fields with sufficient air drainage may escape frost at the beginning or end of the season, resulting in a longer growing period. A heavy soil with good moisture-retaining properties may be preferred in a relatively dry, non-irrigated situation. However, very heavy soils that are wet for extended periods or those compacted by large machinery should be avoided, since adequate soil aeration is needed.

Cucurbits will not tolerate 'wet feet'; if their roots are submerged for too long, the plants will be wilted and stunted, and may not fully recover, even when drainage is later improved. In experiments on summer squash, waterlogging reduced the number and length of lateral roots, but promoted the development of adventitious roots at the stem base near the soil surface (Huang *et al.*, 1995). Salinity in the water exacerbated the negative impacts of waterlogging on root, shoot, leaf and fruit growth. When 14 days of waterlogging were followed by 7 days of drained conditions, the squash plants were able to recover from most of the adverse root effects.

Most cucurbits should not be planted on highly acidic soils. Soil pH of 5.6–6.8 is recommended for cucumber, squash, loofah, chayote and wax gourd. Watermelon tolerates pH as low as 5.0, but grows best at pH 6.0–7.0. Soil pH 6.0–6.7 is recommended for melon plants; they grow poorly and may become chlorotic when grown in very acidic soils. If liming is required, it should be applied long before planting cucurbits to give soil acidity time to adjust.

FERTILIZER

Concerns in industrial countries over water pollution by fertilizer and manure have led to a renewed interest in crop rotation with soil-improving, nitrogen-fixing legumes. Singogo *et al.* (1996) found that fruit yields of the melon 'Magnum 45', grown after a single winter cover crop of alfalfa or Austrian winter pea, were similar to yields produced with application of 100 or 135 kg ha^{-1} inorganic nitrogen. Furthermore, the addition of feedlot beef manure to the soil before ploughing under the cover crop in spring produced only slight increases in the yield of the summer melon crop.

It has long been the practice in Egypt and India to apply manure in trenches before filling the trench with soil and planting watermelon. In most areas today, a complete fertilizer with nitrogen, phosphorus and potassium (N:P:K) is generally ploughed down before planting cucurbits. The type and amount of fertilizer varies in different growing areas. Pierce (1987) recommends 22–34 metric tonnes per hectare of animal manure or green manure equivalent for cucumber. He reported that application amounts of inorganic fertilizer for cucumber ranged from 56 kg ha^{-1} each of N, phosphoric anhydride (P$_2$O$_5$), and potash (K$_2$O) in northeastern states of the USA, to 168 kg nitrate, 134 kg phosphate and 202 kg potash per hectare in Florida. All or part of this fertilizer is ploughed down, and the rest banded at planting.

Scientists tested fertilizer formulas for growing fluted pumpkin in acidic, sandy soil in Nigeria (Obiagwu and Odiaka, 1995). They found that phosphorus was the most important component, and that P alone at 22 kg ha^{-1} gave higher fresh-matter yields than the same amount of P along with N or K. In their experiments, band application or broadcasting were significantly more effective than ring application or point placement methods.

Sidedressing with supplementary N may increase yield, especially for cucurbits having a long growing season, but it is not recommended for cucumber plants which are to be harvested once by machine. Furthermore, ammonium nutrition is detrimental to cucumber plant growth at soil pH <5.0 or >7.5. Hollow heart, in which an empty cavity develops in the fruit flesh, and excessive vine growth of watermelon can be caused by large amounts of N fertilizer. In situations where supplementary N is desirable, applications should be made before flowering, until the vines become too large to permit a tractor between the rows.

Supplementary N and other nutrients can be provided through the irrigation water (a process called fertigation) if plastic mulch or thick vine growth prevent sidedressing close to the plants. Fertigation allows the use of smaller plant beds than those used with band application. When applied via irrigation, a smaller amount of fertilizer can produce the same or greater yields as produced with a larger amount of dry, granular fertilizer (Fig. 5.10). Swiader et al. (1994) found that sprinkler-applied 112N:112K split into five fertigations during the growing season (and supplemented with a preplant application of dry 28N:56K) yielded large numbers of marketable pumpkins of Cucurbita moschata without compromising early maturity. Increasing potassium to 224 kg ha^{-1} produced favourable results as well. However, lower amounts of nitrogen (e.g. 56N:112K) or higher amounts of either nutrient (e.g. 168N:224K) gave smaller fruit yields as a result of later flowering and, in the case of limited nitrogen, poor growth and lower overall flower production (Table 5.7).

Scientists in Florida conducted trials on N requirements for slicing cucumbers grown with polyethylene mulch and drip irrigation (Hochmuth and Hochmuth, 1991). Testing treatments of 0–222 kg ha^{-1}, they found that

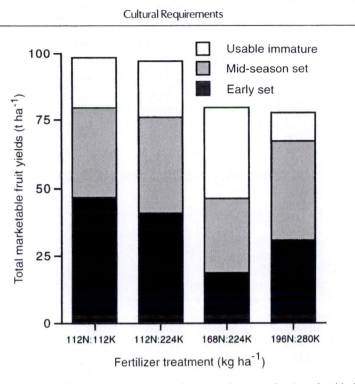

Fig. 5.10. Effect of N:K fertilizer regime on pumpkin (*Cucurbita moschata*) marketable fruiting response in a 1988 trial. All treatments were applied by fertigation except for 196N:280K, which was a dry-blend fertilizer applied before planting. Early set fruits were mature 64 days after seeding (DAS); mid-season set fruits set after 64 DAS, but were mature at harvest; and usable immature fruits were immature at harvest, but >15 cm in diameter, making them suitable for mechanical harvest and processing. (Data from Swaider *et al.*, 1994.)

total fruit yields levelled off with the application of 153 kg ha^{-1} in 1988 (under conditions of leaching rainfall) and 64 kg ha^{-1} in 1989, when conditions were considerably drier.

In glasshouse trials with melon, Sharples and Foster (1958) found that plant dry weight was highest at about 0.26 grams of added ammonium nitrate per kilogram of local soil. At that level of N, additions of mono-ammonium phosphorus and potassium sulphate had little effect on growth. The scientists went on to test the effects and interactions of these nutrients with calcium and magnesium. In the presence of high available N (0.66 g added per kg soil), high Ca content in the soil created a Mg deficiency (<1%) in the leaves; calcium ions also competed with potassium ions to restrict K concentration in leaves (Fig. 5.11). When available N was decreased, low N and high Ca interacted significantly to reduce plant growth and Mg levels in leaves.

Adams *et al.* (1992) tested the interactions of N, K and Mg fed to

Table 5.7. Effect of N:K fertilizer regime on pumpkin (*Cucurbita moschata*) female flowering, per cent fruit set, and plant dry weight at 72 days after seeding.

N:K fertilizer[1] (kg ha^{-1})	Number of days to first female flower	Number of female flowers per plant per day[2]	Female flower fruit set (%)	Plant dry weight (g plant^{-1})
1987 Trial				
56:112	60.0a[3]	0.07b	40.3b	100c
112:56	51.8b	0.12a	53.5a	241b
112:112	51.0b	0.13a	51.9a	271b
196:280	53.0b	0.09ab	57.8a	357a
1988 Trial				
112:112	49.0b	0.14a	53.3a	316b
112:224	50.0b	0.13a	52.8a	339b
168:224	56.0a	0.13a	58.4a	420a
196:280	52.8ab	0.09b	60.2a	481a

[1]All fertilizer treatments were applied by fertigation except for 196N:280K, which was a dry-blend fertilizer applied before planting.
[2]For the period from initial anthesis to 72 days after seeding.
[3]For each year, mean separation within columns by LSD at *P*=0.05.
Modified from Swiader *et al.* (1994), with permission from the American Society for Horticultural Science, 600 Cameron Street, Alexandria, VA 22314, USA.

glasshouse cucumber plants grown in sphagnum peat. As expected, increasing amounts of N or K caused increasing Mg deficiency and corresponding losses in fruit yield.

Mineral element deficiencies, including S, Fe, Mn, B, Cu, Zn, Si and Mo, can cause chlorosis (Fig. 5.12), stunting, poor fruit shape and other problems. For example, insufficient Ca promotes the development of blossom end rot in cucurbit fruits under moisture stress. The first symptom of Mg deficiency is the yellowing of the margins of older leaves. Inadequate boron leads to an increase in auxin activity in squash roots, inhibiting their growth. Silicon deficiency in cucumbers reduces vegetative growth and fruit production, and increases susceptibility to some mildew diseases. Symptoms of various element deficiencies in cucumber are described and illustrated by Rordan van Eysinga and Smilde (1969).

Foliar sprays are an effective means of delivering most minerals. For example, a low-volume spray with 10% magnesium sulfate is suggested for cucumber plants showing signs of Mg depletion. Alternatively, Mg can be added to the nutrient solution (ca 50 ppm) or soil (ca 50 kg ha^{-1}). Melon plants grown on sandy or acidic soil respond to Mg provided in dolomitic lime.

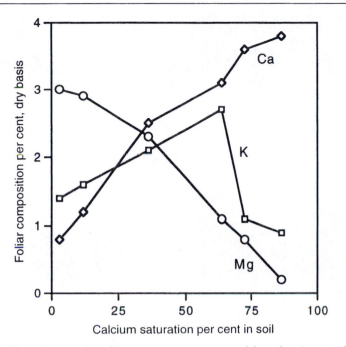

Fig. 5.11. Effect of varying the calcium saturation percentage of the soil on Ca, K, and Mg composition in leaves of melon line CBR-1. (Data from Sharples and Foster, 1958.)

Excessive concentrations of Mg, K, Ca, S and Mn stunt plant growth (Fig. 5.13). Manganese toxicity is sometimes a problem in freshly steamed glasshouse soil; leaf symptoms in melon include pinhole lesions and water-soaked margins on the lower surface.

When nutrients are applied with irrigation water, pH and electrical conductivity (EC) of the solution affect the availability and toxicity of the nutrients to plant roots. For example, Fe, Mn, Co and Zn become less available as alkalinity increases above pH 6.5, whereas Mo, S and K become less available as acidity increases below pH 6.0. Mg and Ca are most available to plant uptake at pH 7.0–8.5.

For adequate cucumber nutrition, the total concentration of elements in solution should be around 1300 ppm, giving an EC reading of about 2.0 dS m^{-1}. Although higher EC (2.5–3.0) can produce stronger plants and better fruit colour in low light situations in glasshouses, values above 4.0 may suppress growth.

Genetic differences in melon for salt tolerance have been reported, but no cucurbit grows well in highly saline media. High total salts content (especially sodium chloride) restricts uptake of certain elements (e.g. Fe), causes mineral toxicities (e.g. Na, Cl, B) and increases tolerance to other elements, such as Zn. Secondary effects are inhibited root development, reduced growth, dark green

Fig. 5.12. These cucumber leaves exhibit interveinal chlorosis as a result of severe iron deficiency.

Fig. 5.13. Cultivar differences in response to manganese toxicity. Left: 'Wisconsin SMR 18', right: 'Ashley'.

foliage with a marginal necrosis and lower yield. When summer squash plants were watered with saline solution (100 mmol m^{-3} NaCl) for 21 days, number of fruits per plant was 45% lower than for plants grown under non-saline conditions (Huang *et al.*, 1995). In melon, increasing salinity in the EC range of 1.2–6.0 caused fruit yields to decrease as a result of decreasing fruit size (Mendlinger, 1994). However, increasing salinity also produced higher percentages of soluble solids in fruits. Similarly, cucumber plants grown in saline conditions of about 3000 ppm salts may yield 25% less than normal, but produce sweeter tasting fruits.

IRRIGATION

Cucurbits, with their long vines and large leaves, have high rates of water loss via transpiration. Consequently, they often benefit from supplementary irrigation, especially when grown on light, sandy soils. Field-grown cucumber plants may need water every 5–7 days and are usually irrigated, even in humid areas. Heavy soils that have been soaked just before planting to a depth of 2 m may contain sufficient moisture for average crops of melon, squash and watermelon without the need for additional irrigation. Vining cultivars of squash and pumpkin may wilt on hot, dry afternoons if not irrigated, but recover turgidity the next morning. Watermelon and wax gourd have deep roots and can survive relatively dry conditions. However, watermelons and other cucurbit fruits can develop a blossom end rot during periods of moisture stress. Early evidence of serious water stress in cucurbits includes the loss of colour in the lobes of older leaves and the tips of developing fruits, and irreversible wilting.

Water stress affects all major physiological processes, from photosynthesis to carbohydrate metabolism. It also causes the hydrolysis of protein. In pumpkin, wilted plants exhibit large increases in free proline, which appears to function as a storage compound during times of stress.

Cucurbits are usually grown on raised beds if furrow irrigated, and on level soil with other irrigation practices. Raised beds dry out faster than flat rows and require more frequent watering. In the USA, commercial growers of field crops often use mobile overhead sprinklers. The supporting equipment may be self-propelled and the sprinklers slowly rotating to ensure total field coverage. In flat fields with a high water table, irrigation water which is added to the soil through ditches or pipes becomes available to plant roots via capillary action in the soil. This seepage system reduces fertilizer leaching. Drip or trickle irrigation has become popular for cucurbits in Israel and other areas where water conservation is important. Water slowly emanates from emitters on plastic pipe directly to the root zone of cucurbits, with little water loss by runoff or evaporation. Compared with furrow, seepage or sprinkler irrigation, drip irrigation conserves water and helps prevent accumulation of soluble salts in the soil. It may also result in fewer foliar diseases than overhead irrigation practices, which leave free water on the leaves.

To prevent fruit rot, irrigation should be avoided as fruits near maturity. Sudden irrigation or flooding following a period of drought can cause fruits to crack, making them more vulnerable to pests and pathogens. Melons of the Momordica Group are very prone to cracking, which usually happens near fruit maturity. Excessive soil moisture also causes plant wilting and stunting.

Glasshouse-grown cucumbers should be thoroughly watered the first few days after transplanting, then irrigated less for about 2 weeks to promote healthy root formation. Thereafter, a regular irrigation schedule can be adopted, e.g. 15 min of drip irrigation every 2 h from dawn to dusk in the summer, depending on temperature. Irrigation should be performed during the day, when transpiration rates are high. Water or solution application should not be so great or of such long duration that drainage is impeded, restricting root access to oxygen. Irrigation scheduling can be managed with the use of a tensiometer to monitor water potential of the soil or other medium.

Pier and Doerge (1995) evaluated the combined effects of soil water tension and N fertilizer on trickle-irrigated watermelon plants grown in Arizona fields. Predicted maximum marketable fruit yield was 102 Mg ha^{-1} at 7.2 kPa tension (measured at 0.3 m depth) and 336 kg N ha^{-1} (supplied as urea via fertigation in five applications). However, this combination of water and fertilizer inputs created unaccounted for, or 'lost', N estimated at about 60 kg ha^{-1}, an amount that could cause unacceptable contamination of ground water. When unaccounted for N loss was limited to <15 kg ha^{-1} and the costs of water and fertilizer were taken into account, 95% of the maximum yield and net return values occurred with an average soil water tension of 7–17 kPa and N applications totalling 60–315 kg ha^{-1}.

SUPPORT, TRAINING AND PRUNING

Field-grown cucumber plants are often allowed to grow freely on the ground, whereas glasshouse cucumbers are usually trained vertically or at an angle to conserve space. Cucumber plants in the field are sometimes grown on trellises for the same reason. Chayote, bottle gourd, loofah, bitter melon, small-fruited wax gourd cultivars and other minor cultigens with large, climbing vines (e.g. species of *Trichosanthes*) are commonly grown on outdoor trellises, particularly in Asian countries.

A trellising system can be made of long strips of fencing, or be as simple as three leaning poles joined together at their apices. In a system composed of vertical support posts and a single horizontal top wire or pole, growers suspend strings from the horizontal support to help train the plants to the top of the trellis. Cucurbits with large, heavy fruits (e.g. wax gourd or large-fruited bottle gourd cultivars) should be supported by very sturdy trellises.

In the glasshouse, a V-cordon system trains cucumber plants to grow upwards at an angle in order to make best use of light. Two support wires positioned up to 1 m apart are anchored about 2 m above a row of plants. Plastic strings lead some branches up to one support wire and other branches to the other wire. Plastic clips secured to the string are positioned every 30 cm or so along the branches under leaf petioles for vine support.

The renewal umbrella system trains glasshouse cucumbers vertically until they reach and are secured to the support wire. Then, the growing tip of the main stem is cut to encourage lateral branching. All lateral branches are removed except for two at the top of the plant, which are trained over the support wire and then down along opposite sides of the main stem. Cucumbers are allowed to develop along the top third of the main stem and on the two laterals. As these fruits mature, two more laterals are allowed to grow from the top of the plant; all other new laterals are removed. When fruits are ready for harvest, the branches on which they occur are cut and new fruits are allowed to develop on the next set of hanging laterals. This process can be repeated over a 10-month period, with single plants producing more than 100 fruits (Resh, 1987).

Pruning vegetative parts and flowers can affect female flower earliness and the quality of fruits on a plant. Terminal shoots on bottle gourd plants should be cut off when they are about 3 m long to promote lateral branching and female flower production. Topping the main stem increases sponge length and quality in fruits of smooth loofah (Table 5.4), whereas pruning laterals promotes flowering. Melon plants grown in cloches are often pruned to keep the plants within the restricted growing area and to remove excess fruit. However, leaf pruning of melon plants after fruit set has begun can impair yield and fruit quality.

If too many fruits are allowed to develop per plant, quality may be adversely affected. For glasshouse melon cultivars, leaves, lateral branches and fruits are sometimes pruned so that only one fruit per lateral and a total of four fruits per plant are maintained. Some watermelon growers remove all but two fruits per plant in order to increase fruit size and uniformity. The superfluous fruits should be removed when still quite small. More fruits per plant can be left on the vines of small-fruited watermelon cultivars. Squash and pumpkin growers striving to produce fruits of enormous size permit only one fruit to develop on a plant.

WEED CONTROL

Weed control is essential for good yield of cucurbits, and it makes harvesting easier. Cultivation, hoeing and other mechanical means of tillage are used. Deep cultivation should be avoided since cucurbits have many shallow roots.

Once the vines are well-established and cover the ground, their abundant foliage canopy intercepts light and inhibits growth of late-germinating weeds.

Genetic differences in allelopathy have been found and proposed as a means of weed control for cucurbits. Roots of some lines of cucumber reportedly release chemicals that inhibit germination of weed seed in the surrounding soil. This characteristic has not been bred into any cultivar, however, and is not used commercially to aid weed control.

Black plastic and infrared-transmitting mulches are often used to prevent weeds from growing. (See Mulches and plant covers, this chapter, page 124 for more details.)

Cucurbits generally are very sensitive to some herbicides, although a few cultivars may be more tolerant. For example, they can be injured by atrazine residues in the field from a previous corn crop, or by 3,6-dichloropicolinic acid residues in straw bale propagation systems. Although ethalfluralin is currently registered for cucurbits in the USA, misuse of this and other dinitroaniline herbicides is a common cause of injury to cucurbits (Zitter *et al.*, 1996). Wet, cool conditions may increase the chance of injury from these and some other herbicides (e.g. chloramben). A herbicide injurious to cucurbits, such as glyphosate, can be applied to the soil between strips of plastic mulch if directed away from the cucurbit plants.

There are other herbicides on the market that can be used safely in cucurbit fields. For example, pumpkin plants are tolerant of the pigment inhibitor clomazone. Most herbicides are applied before planting, but the cell division inhibitor naptalam can be utilized after the last cultivation. Sufficient soil moisture is required for various of these herbicides to be effective. Better, more selective herbicides are needed for cucurbits.

The 'stale seed bed' technique has been employed to prevent weeds from becoming a problem in cucurbit fields. Several weeks before the cucurbits are to be sown, the field is ploughed and prepared for planting. This encourages weed seeds to germinate, and then the weeds are killed with a non-residual herbicide (e.g. glyphosate) before the cucurbit seeds are planted. When glyphosate is used for control, it should be applied to weeds 3 days before cucurbit seeding.

Soil fumigation is very effective for weed control, and it also controls soilborne diseases and nematodes. It is seldom employed for cucurbits in the field because of the cost, but has been used for cucumber plants grown in glasshouse ground beds.

6

FRUIT AND SEED PRODUCTION

CONTROLLING SEX EXPRESSION

Although some summer squash cultivars can set fruit without pollination, many cannot. Poor fruit set in the latter group is a common complaint early in the flowering season, when there may be insufficient flowers of both sexes for reliable pollination. Seedsmen have selected summer squash lines that produce many pistillate and few staminate flowers. This can contribute to early fruit set and has reduced the cost of producing hybrid seed in cases where emasculation is accomplished by manually removing male flowers from the female parent. However, it has created problems when predominantly female cultivars are planted early in the season, because low temperatures promote female sex expression even more, to such a degree that male flowers for pollination may be lacking. This problem can be overcome by interplanting a highly pistillate cultivar with a predominantly male cultivar (e.g. 'White Bush Scallop'), but this is seldom done. Instead, most growers simply wait until the cultivar planted produces both staminate and pistillate blossoms.

The use of gynoecy to produce hybrid seed without the need for emasculation has become extremely important for cucumber, though not for melon. In fact, gynoecious cultivars represent one of the most important developments of cucumber breeding in modern times. Since the gynoecious inbred parent of a hybrid cultivar generally has only female flowers, the open-pollinated seeds it produces are F_1 hybrids. Thus, costly hand pollination and emasculation are not necessary.

Most gynoecious hybrid cucumber cultivars are heterozygous for F, the allele for female sex expression. Under some environmental conditions, these cultivars produce male as well as female flowers. Cultivars stable in gynoecious sex expression can be bred by selecting for the right combination of modifying genes with F/F^+, or for the homozygote (F/F). Homozygous gynoecious hybrid seed has been created by hand pollinating two gynoecious

145

inbreds in isolation, usually in a glasshouse, after treating the male parent with a growth regulator to induce male flower formation. However, this method of producing homozygous seed is expensive. A more efficient method is to use a hermaphroditic line (which has the genetic background $F/F\ m/m$) as the male parent for the hybrid. The parental lines are grown in adjoining rows, and cross-pollination is accomplished by bees instead of by hand. Hybrid seed is harvested only from the gynoecious inbred ($F/F\ m^+/m^+$) used as the female parent. The hybrid is homozygous gynoecious because both parents were homozygous for the gynoecious allele F. Its being heterozygous for the andromonoecious allele m is of no consequence since this allele is recessive and has no effect on the hybrid.

If a gynoecious cucumber hybrid is not parthenocarpic, its seed is usually blended with 10–15% of seed from a monoecious cultivar to ensure sufficient male flowers for pollination. 'Sumter' has been used as a pollenizer for single-harvest, gynoecious field cultivars; its first staminate flower opens about 35 days after planting. Recently, Walters and Wehner (1994) found several sources of breeding material in which anthesis begins within 26–30 days after planting.

In some cucurbits (e.g. bottle gourd), removing male flower buds will increase vegetative growth, including the production of lateral branches, where a greater percentage of female flowers occur. However, this method of promoting femaleness is not effective in cucumber (Delesalle and Mooreside, 1995).

Many growth regulators have been reported to influence sex expression. Ethylene is a naturally occurring plant hormone which plays an important role in this and many other physiological processes in cucurbits. The effects of light intensity, photoperiod and other environmental factors on sex expression may be mediated through their influence on endogenous ethylene.

Ethylene is generally associated with femaleness. For example, gynoecious cucumber plants generate more ethylene than monoecious plants. Exogenous application of ethylene, or compounds that release ethylene, can induce a bisexual floral bud to develop into a pistillate flower in cucumber, melon and squash. However, ethylene has been shown to promote maleness in watermelon (Rudich, 1990).

Currently, the growth regulator most widely used to stimulate production of pistillate flowers is ethephon. The active ingredient of this growth regulator, 2-chloroethylphosphonic acid, is an unstable compound. When its pH increases, such as when it is applied to cucurbits, this compound breaks down to release ethylene. Monoecious cucumber and squash plants treated in the seedling stage with ethephon produce only pistillate flowers for an extended time and have reduced internode lengths (Fig. 6.1; Robinson et al., 1969, 1970). However, ethephon treatment of smooth loofah seedlings caused no change in sex expression in recent experiments (Wehner and Ellington, 1995).

Fig. 6.1. Left: untreated monoecious zucchini plant; right: zucchini plant treated with ethephon, resulting in gynoecy. Leaves of both plants have been removed to show flowers. (Robinson *et al.*, 1970. Reprinted courtesy of Kluwer Academic Publishers.)

Ethephon is widely used to facilitate production of F_1 hybrid summer squash seed. This seed was formerly produced by hand pollination or by manually removing staminate flower buds from the inbred line used as the maternal parent, with bees bringing pollen from a male parent grown in an adjoining row. Now, ethephon eliminates the need for laborious removal of staminate flower buds to prevent selfing or sibbing of the maternal parent. Several rows of the ethephon-treated female parent are usually grown for each row of the male parent in seed production fields. After fruit set is completed, plants of the male parent are eliminated from the field so that their seeds will not become mixed with those of the F_1 hybrid.

Squash lines differ in response to ethephon. Plants with a high degree of female sex expression, i.e. having a high ratio of pistillate to staminate flowers when untreated, are generally easy to convert to gynoecy. Several applications of the chemical are usually required during the flowering season to prevent the female parent from producing male flowers. Shannon and Robinson (1979) determined that two applications of 400–600 µl l⁻¹ significantly inhibited staminate flower development in summer squash without reducing seed yield or quality.

Ethephon has also been applied to the female parent of hybrid cucumber cultivars that are monoecious, but there is no need for it with gynoecious hybrid cultivars unless the female parent produces some staminate flowers.

In order to maintain a gynoecious cucumber line by selfing or sibbing, it can be induced to develop male flowers with the application of another important endogenous regulator of sex expression, gibberellic acid (GA). GA has the opposite effect of ethephon: it typically stimulates development of staminate blossoms, although it has been shown to promote femaleness in bitter melon. GA_{4+7} is much more effective than GA_3 for stimulating the production of staminate flowers in gynoecious cucumber plants. Also used for this purpose are silver compounds, such as silver nitrate or silver thiosulphate. Silver compounds, aminoethoxyvinyl glycine (AVG), and CO_2 promote maleness by inhibiting the activity of ethylene.

Male-sterile mutants have been found for cucumber, melon, watermelon and squash. In each case, sterility is inherited as a single recessive allele. Since a cytoplasmic factor is not involved, lines that breed true for this character are not possible. This is a limiting factor in the use of male sterility to produce seed, since the female parent will segregate for male sterility and male-fertile plants must be rogued in seed production fields. All male-fertile plants should be removed from the female parental line before flowering, but this can be difficult since they cannot be recognized until the male flower buds form a few days before anthesis. Glabrous foliage is associated with a male-sterile allele (*gms*) of watermelon, making it possible to classify plants for this allele in the seedling stage. Male sterility has been used to a limited extent in the production of F_1 seed of the squash *Cucurbita maxima*. Although not extensively used, male sterility is potentially advantageous for producing melon seed, since it is tedious and expensive to remove the anthers from hermaphroditic flowers of the andromonoecious maternal parent of a hybrid. Male sterility can be induced by high concentrations (>2.2 mmol l^{-1}) of maleic hydrazide.

Other compounds that can be applied to suppress maleness or promote femaleness are auxins, 6-benzylaminopurine, flurenol, chlorflurenol, abscisic acid and boron. In melon, flurenol or chlorflurenol is applied to the foliage of young plants or the seeds are soaked in the solution for 4–5 days (Roy and Saran, 1990). Various growth regulators (e.g. 6-benzylaminopurine) can also be introduced through the roots of seedlings.

POLLINATION

Flowers of cucumber, melon, squash and watermelon open in the early morning, and pollination is generally completed by noon. Blossoms of some white-flowered species (e.g. bottle gourd) open at night, but remain at anthesis the next morning.

Most species of the *Cucurbitaceae* are pollinated by bees or other insects, which are attracted to pollen and nectar (Fig. 6.2). Different species of

Peponapis and *Xenoglossa* (gourd bees) preferentially pollinate different species of squash. Cucumber beetles can pollinate bottle gourd and other cucurbits. Pollination by hummingbirds and butterflies has been reported for *Gurania* and *Psiguria*, and moths and bats are believed to pollinate cucurbits with white, night-opening flowers.

Some growers place hives of honey bees in their cucurbit fields when the plants flower in order to improve fruit set; this gives higher quality fruits or better seed production, depending on crop objectives. Two or three large bee colonies per hectare, positioned within 100 m of the edge of the field, are recommended for cucumber. For melon, the bee colonies should be placed in the field, as opposed to on the borders, after flowering has begun, or the bees might migrate to other fields. Care should be taken when applying pesticides, to avoid killing pollinators. Insecticides with short residual activity are best,

Fig. 6.2. Insect pollinator visiting a male flower of wax gourd.

and should be applied in the afternoon since pollinating insects are usually in cucurbit fields in the morning.

Bees can bring cucurbit pollen from great distances, and cross-pollination with a bitter ornamental gourd (*Cucurbita pepo*) is possible even in isolated seed production fields of that species of squash. An alert seedsman will recognize the crosses the next generation, since hybrids between summer squash and an ornamental gourd will generally have shorter, rounder fruits with warts and different rind colour than is typical for the squash cultivar being grown. Even though the offending hybrid plants are rogued and not allowed to produce seed, bitterness can persist in succeeding generations if bees distributed pollen from the gourd-squash hybrid plants before the plants were removed from the field.

As part of their quality control programme, it is recommended that seedsmen self-pollinate a number of individual plants of open-pollinated squash cultivars or the inbred parents of F_1 hybrid cultivars. After the selfs are grown out and tested for bitterness, remnant seeds of all selfs proved to not carry the allele for bitter fruit can be combined as elite seed stock for future increases of the cultivar.

Cucurbits are self-compatible. Reports of self-incompatibility in squash have been proved to be spurious. Most monoecious cucurbits are naturally self- as well as cross-pollinating.

The bisexual flowers of andromonoecious melon and cucumber plants seldom set fruit if not mechanically pollinated. Although they are self-compatible and will set fruit if self-pollinated, they normally require the assistance of a pollinator because pollen is shed towards the outside of the flower, away from the centrally-located stigma. When pollinated by insects, hermaphrodite flowers generally have a higher rate of successful self-pollination than do pistillate flowers.

Inadequate pollination can cause poor fruit shape in cucumber. Although pollen normally takes a few hours to reach ovules, if the rate of fruit elongation exceeds the rate of pollen tube growth, then the more distant ovules in long-fruited cucumbers are never fertilized. Also, if there are insufficient pollen grains on the stigma, generally only the ovules closest to the stigma are fertilized. In either case, the blossom end of the fruit is stimulated by fertilization to enlarge more than parts of the cucumber without seeds, resulting in a misshapen fruit.

'Seedless' cucumber cultivars will produce seeds if pollinated; hence, seed production for parthenocarpic cucumbers is similar to other types. Seeds produced per fruit may be low if the ovary is very long, due to the distance pollen tubes must travel to fertilize distant ovules. For fruit production, glasshouse vents can be screened to exclude insects which might pollinate and cause poor fruit shape in seedless cucumbers.

Parthenocarpy can sometimes be induced by rubbing the stigmas with pollen of another species. In cucumber, chlorflurenol (2-chloro-flurenol-

9-carbonic acid) is very effective for promoting parthenocarpy (Fig. 6.3; Robinson *et al.*, 1971). Gibberellins applied to the stigma can cause seedless parthenocarpic fruit development in chayote (Aung *et al.*, 1990). A newly developed cytokinin, 1-(2-chloro-4-pyridyl)-3-phenylurea, which promotes fruit set in pollinated flowers of melon and watermelon, also produces parthenocarpy in unpollinated flowers of watermelon (Hayata *et al.*, 1995).

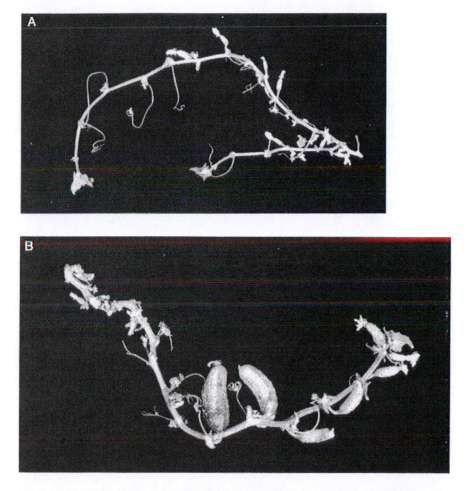

Fig. 6.3. (A) Two applications of 250 ppm ethephon produced gynoecy in this cucumber plant. (B) When chlorflurenol was applied to a similarly treated plant, its fruits developed parthenocarpically. (Reprinted from Cantliffe *et al.*, 1972, with permission from the American Society for Horticultural Science, 600 Cameron Street, Alexandria, VA 22314, USA.)

HARVESTING

Minor cucurbit crops, which are grown primarily in non-industrialized countries, are harvested by hand. Two to three months after planting, edible immature fruits of bitter melon, chayote, wax gourd, angled loofah and bottle gourd can be picked weekly throughout the fruiting season. In temperate regions, smooth loofahs grown for sponges, bottle gourds grown for containers and wax gourds grown for storage generally should remain on the vine until the end of the growing season. In tropical areas with year-round growing, smooth loofahs are mature when the rind turns brown and the seeds rattle inside the lightweight, dry fruits. Bottle gourds can take 4–6 months to mature on the vine, depending on the cultivar; the fruit turns from green to light brown or yellow when mature. Those harvested too soon may have thin rinds and not store well.

Multiple harvests are made by hand for melons, and are sometimes performed at night if the days are hot. Melons harvested when immature have low sugar content and poor flavour. However, Conomon Group melons grown for pickling should be picked when immature, once they are of full size.

Maturity of cultivars of the Cantalupensis Group can be gauged by the appearance of the 'slip', the abscission zone on the peduncle at its attachment to the fruit. A melon is said to be 'full slip' when the abscission layer forms a ring entirely circumscribing the fruit attachment, causing the ripe fruit to detach naturally, or slip, from the vine. Fruits harvested before half slip, when the ring is no more than half completed, may be inferior in quality. They can be recognized by the jagged scar at the stem end of the fruit where it was detached before completion of the abscission layer. For 'PMR 45', half slip occurs 35–40 days after anthesis.

Inodorus Group melons do not form an abscission layer. Their maturity is judged, with some difficulty, by the size, shape, feel, and aroma of the fruit. For example, a 'Casaba' melon is deemed ready for harvest when there is a slight yielding at the blossom end when thumb pressure is applied. Although 'Honey Dew' melons can be harvested as soon as 35 days after anthesis, ethylene treatment is usually needed to ripen the fruits sufficiently for market. When left on the vine, the fruits naturally become mature around 50 days after anthesis, which is too late for commercial harvesting since over-ripening occurs within a few days (Pratt *et al.*, 1977).

Watermelon fruits do not naturally separate from the vine, and it may be difficult for the inexperienced to judge maturity. Size and colour of the fruit changes as maturity approaches, with the fruit becoming larger and fuller and the ground spot, where the fruit rested on the ground, changing from white to light yellow. The tendril closest to the fruit may wither and turn brown when the fruit ripens, but this is not always a reliable indicator. The time-honoured ritual of thumping the fruit is done to determine if the sound is sharp, ringing or metallic, indicating immaturity, or muffled and resonate when the fruit is ripe. Watermelons are of highest quality when harvest is delayed until the fruit has 10% or higher soluble solids content.

Watermelons should not be harvested in the early morning, when they are more turgid and apt to crack. A sharp knife is used to cut the fruit from the vine, leaving about 3 cm of peduncle attached to the fruit to deter stem end rot. The peduncle may be treated with a paste or wax to further limit microbial access.

Winter squashes and pumpkins should be hand-harvested before freezing weather damages the fruit and reduces storage life. A light frost will not hurt mature fruits, and may facilitate harvesting by killing the vine cover and exposing the fruits, but prolonged exposure to temperatures below 10°C causes chilling injury. In tropical countries, mature squashes are picked when the rind loses its sheen and the tendril nearest the fruit dies. As with watermelons, 2–3 cm of peduncle is often left on the fruit when packing into small cartons or crates. For direct harvest into large pallet bins, peduncle length should be reduced, or the peduncle removed, to avoid damage to other fruits.

Cucumber, summer squash, bitter melon and other cucurbits grown for their immature fruits are more productive if the fruits are harvested frequently (e.g. every other day) and not allowed to become too large. Large fruits act as a sink for nutrients and inhibit development of additional fruits on the plant. Summer squashes should be picked soon after the blossom falls off; they are more tender and slightly sweet at this stage. In China, the blossom may be left on slicing cucumbers to indicate freshness of the young fruit. In the USA, pickling cucumbers are sold on the basis of size, the smallest fruits having the greatest value. The proportion of fruits in small size grades is increased if harvesting is performed at frequent intervals.

Although slicing cucumbers are harvested by hand, mechanical harvesters have been developed for pickling cultivars (Fig. 6.4). Some harvesters are designed to make multiple harvests, but the most popular harvesters make only a single harvest in a field. It is important for a once-over, mechanical harvest that the cultivar and cultural conditions be such that the maximum number of fruits are ready for harvest at the same time. Gynoecious cultivars with uniform germination are recommended. Planting density should be at least 100,000 plants per hectare. Price and Kretchman (unpublished, 1990) recommend that the plants be grown on well-drained, uniform soil with less than 1% slope and free of large stones. Long fields are efficient, since they minimize the time spent turning the harvester at the end of the row. Fields with a persistent herbicide residue or heavily infested with weeds should be avoided. Rotation is recommended to avoid planting in areas where cucumber or pepper crops were infected with *Phytophthora* the previous year, or in soils with residues from the previous corn crop. To increase yield, parthenocarpic fruit set can be induced by the growth regulator chlorflurenol, where governmental regulations permit its use.

Fig. 6.4. Mechanical harvester of pickling cucumbers in California.

For once-over harvesting, prediction of harvest date is important and has typically been based on the number of days from planting to harvest of previous years' crops. Perry and Wehner (1990) tested a heat unit model labelled 'reduced ceiling' for predicting optimum harvest date for pickling and slicing cultivars of cucumber. This model sums, over days from planting to harvest, the difference between the daily maximum (except if the maximum is greater than 32°C, in which case it is replaced by 32°C minus the difference between the maximum and 32°C) and a base temperature of 15.5°C. They found that their heat unit summation value was a better predictor of harvest date than the standard method of counting days for picklers, but not for slicers.

Harvest aids (e.g. tractor-drawn conveyer belts) have been used to improve efficiency in collecting and processing cucumbers, melons, water-melons and other cucurbits grown in large fields. The harvest aid closely follows the pickers so that the workers can easily place the fruits on to the conveyer belt, which carries the produce to the packing area (Fig. 6.5).

Fig. 6.5. Cucumbers being transferred by conveyer belt from the harvester into slatted wooden shipping bins.

POSTHARVEST HANDLING AND STORAGE

Fruits

Malabar gourd and wax gourd fruits have the longest shelf-lives among cucurbits grown for consumption. They can be stored for a year or more in a cool, dry environment. Although cucurbit fruits generally keep longer at lower temperatures, they should not be frozen or have prolonged exposure to chilling temperatures (<10°C, exact temperature depending on the cultivar). A change in respiration that leads to the production of harmful end products (e.g. ethylene) at low temperature appears to contribute to chilling injury. In cucumber, pitted rinds and collapsed flesh are evidence of damage. African horned cucumbers develop opaque rind spots when stored at 4°C. Although fruits of this species will keep for 3 months at 24°C, shelf-life shortened at lower temperatures in recent experiments. At 20°C, there was 30% spoilage among stored fruits by day 37 (Fig. 6.6); at storage temperatures below 12°C, all ripe fruits spoiled within 55 days (Benzioni *et al.*, 1993).

Cucurbit fruits harvested immature (e.g. summer squash, cucumber, bitter melon, angled loofah, chayote) should be handled carefully to avoid damaging their thin skins. Rind wounds give fruits an unmarketable appearance, allow entry of pathogens and reduce shelf-life. Unprocessed,

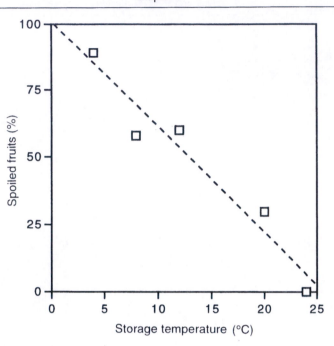

Fig. 6.6. Effect of storage temperature on the percentage of spoiled mature fruits of African horned cucumber after 37 days of storage. (Data from Benzioni *et al.*, 1993.)

immature fruits usually survive storage for only a few weeks. Bitter melons keep for about 2 weeks at 12–13°C and 85–90% relative humidity (RH). Summer squashes can be kept safely at 7–10°C and 85–95% RH, but chilling injury may occur at lower temperatures. Cultivars of *Cucurbita pepo* with allele B (see 'Genes,' Chapter 3, page 46) in their genetic background are more susceptible to chilling injury and shrivelling during storage than similar cultivars lacking this allele. Shrivel symptoms may appear when there is as little as 6% weight loss in fruits of some straightneck squash cultivars (Sherman *et al.*, 1987).

Winter squashes and pumpkins should be fully mature before being harvested for storage. They can be kept in bulk bins, either temporarily before processing (Fig. 6.7) or for long-term storage. Slatted bins allow better air circulation and fewer losses to disease than solid-wall bins. Care should be taken not to overfill storage containers as compression damage to fruits is possible. For extended storage, the bins are loaded onto trucks and moved into ventilated storage buildings by forklift.

If winter squashes and pumpkins are to be held for a long time, those of some cultivars can be cured by being kept for 3–20 days at 20–30°C. The purpose of the warm temperature treatment is to promote healing by callus formation over any wounds acquired during harvest and transit. The curing

Fig. 6.7. Winter squashes (*Cucurbita maxima*) (A) being transported from the field to (B) a canning plant.

process also hastens starch conversion to sugar (Fig. 6.8). However, curing can damage some cultivars; the skin colour, flesh texture and taste of 'Table Queen' fruits are adversely affected by this process.

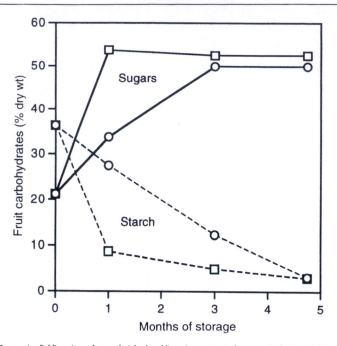

Fig. 6.8. Sugar (solid lines) and starch (dashed lines) content changes in fruits of 'Butternut' (*Cucurbita moschata*) during storage. Fruits were cured at about 26°C for 3 weeks (squares) or not cured (circles). (Data from Schales and Isenberg, 1963.)

Cured and non-cured squashes should be kept cool and relatively dry, preferably at 10–15°C and 50–70% RH. A higher RH of 70–80% will reduce shrinkage. For 'Butternut', weight loss should be kept below 15% to minimize development of a hollow neck. The disadvantage of high humidity storage is that microbial damage is greater. Hot water treatment (2 min dip in 60°C water) has been used to reduce decay in 'Butternut' fruits stored at 70% RH (Francis and Thomson, 1965).

Mature squashes can be stored for up to 6 months, depending on the cultivar; 'Table Queen' and other acorn squashes store well for 2 months, 'Turban' for 3 months, and 'Hubbard' for 6 months. During this time, starch conversion continues and fruit colour and β-carotene content may improve (Fig. 6.9). However, the palatability of 'Hubbard' fruits declines over time as overall carbohydrate content decreases and the percentage of water increases (Cummings and Jenkins, 1925).

After harvest, melons are transported to a packing shed. If they retain much field heat, they are cooled in a hydrocooler, which is kept cold by a combination of ice and mechanical refrigeration. Some melons, particularly 'Crenshaw' and other thin-skinned cultivars, may be packed into padded shipping boxes in the field directly after they are picked. Melons transported

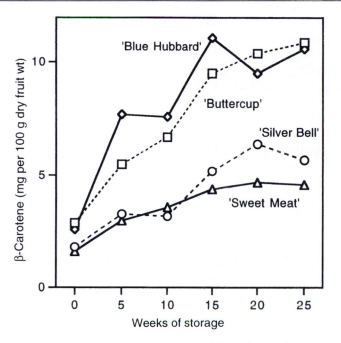

Fig. 6.9. Dry basis β-carotene content in mature fruits of four cultivars of winter squash (*Cucurbita maxima*) during cold storage. (Data from Hopp *et al.*, 1960.)

long distances after harvest are often shipped in iced railroad cars or refrigerated trucks.

Melons can be stored for 1–6 weeks, depending on the cultivar and storage conditions. Cantalupensis Group melons fare well at 3–4°C and 85–90% RH for up to 2 weeks. Inodorus Group cultivars can be held for longer periods at temperatures up to 20°C.

'Honey Dew' melons are frequently treated with ethylene after shipment to promote uniform ripening of the harvest; however, immature fruits will not ripen off the vine satisfactorily, even with the application of ethylene. Late-maturing pumpkins are treated with ethylene (ca 1000 µl l^{-1}for 24 h) to promote orange coloration. Application of 160 µl l^{-1} ethylene for 24 h to African horned cucumbers caused green, nearly mature fruits to turn yellow within 3 days of treatment compared with over 20 days for untreated fruits (Benzioni *et al.*, 1993). Atmospheric ethylene, from stored fruits such as apples, a faulty heater or other source, should be avoided when storing many other cucurbits; it causes undesirable yellowing of cucumbers and green-skinned squashes and flesh softening in watermelon.

Cucumbers should be handled carefully and cooled, in ways similar to melons, as soon as possible after harvest. Rough handling (e.g. dropped, smashed or cut fruits) during harvest or grading can cause bloating. Cultivars vary in their susceptibility to fruit damage; for example, fruits of 'Chipper' are more resistant to skin fracture than those of 'Pioneer'.

In the USA, growers take their crop to commercial receiving stations where cucumbers are sorted by grade and then trucked to pickle factories, where they may be stored before processing. At the pickling plant, cucumbers are: (i) brined, with or without fermentation (which can be controlled), and then stored or processed into dill, sweet or sour pickles; or (ii) packed with a cover liquor consisting of water, vinegar, salt and other flavourings, and then pasteurized to produce 'fresh pack' pickles or kept cold for refrigerated dills. These processes are illustrated in Fig. 6.10 and described in detail in Miller and Wehner (1989). For brined products, brine tanks are purged with air or nitrogen to remove CO_2 in order to reduce fruit bloating. Later, the pickles are desalted by leaching.

Recommended pre-processing storage conditions at pickling plants are 10°C and 95% RH (Etchells et al., 1973). Lower levels of humidity cause fruit moisture loss, and higher temperatures promote microbial growth. Cultivars vary in their storage life; under favourable conditions, 'Ohio MR200' and 'Marketer' will keep for 10 or 47 days, respectively. When recommended storage times are exceeded, mould and bacterial infections take hold (especially in spots where the spines have been broken off), the rind becomes loose on the shrivelling fruit, and flesh flavour and texture deteriorate.

In experiments by Fellers and Pflug (1967), small (<33 mm diameter) and large (33–51 mm) unwashed pickling cucumbers could be stored safely at 4.4°C for 6 and 9 days, respectively. Washing reduced storage life in half, whereas maintaining atmospheric levels of 5% CO_2 and 5% O_2 more than doubled storage life. However, such a low temperature is not usually employed in commercial operations because chilling injury can occur and the fruits deteriorate quickly (within 18 h) after removal from storage.

As with picklers, high humidity (ca 95% RH) slows softening and shrivelling due to water loss in slicing cucumbers during storage. Slicers are waxed after harvest to keep the fruits from drying out as well as to enhance appearance. Glasshouse-grown cucumbers, which have a very tender skin, are often wrapped in plastic to prevent bruising and retard dehydration.

Watermelons should be harvested at maturity, as flavour and soluble solids content generally do not improve with storage. However, mature fruits are more easily injured during postharvest handling. They split easily if dropped and are subject to compression damage. They should be hand-packed in straw- or foam-padded crates to prevent bruising and cracking during shipment.

Watermelons should be cooled as soon as possible after harvest. Delayed cooling and high humidity can lead to activation of anthracnose, a disease that may be latent in the field. Watermelons store satisfactorily at 15°C for up to 2 weeks. For long-term storage, the fruits should be kept at about 12°C and 85% RH; flesh colour fades below 10°C.

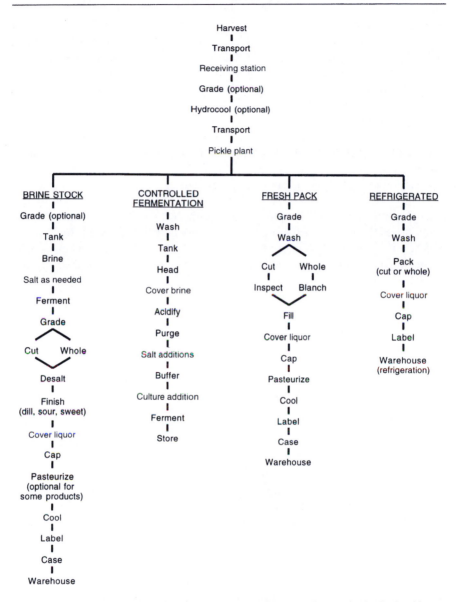

Fig. 6.10. Flow chart showing the movement of cucumbers from harvest to finished pickles. (Redrawn from Miller and Wehner, 1989. Reprinted with permission from N.A.M. Eskin (ed.) *Quality and Preservation of Vegetables.* Copyright CRC Press, Boca Raton, Florida, © 1989.)

After mature bottle gourds are harvested, they need to dry out in a cool, dry, dark area for another 3–9 months, until the seeds rattle inside. Before storing, the freshly picked fruits should be washed thoroughly with a disinfectant solution to remove dirt and mould. During storage, they should be checked for mould weekly and cleaned if necessary. After curing is complete, the gourds are soaked to soften the outer skin, which can then be scraped off with a knife. After allowing the scraped fruit a day to dry, it is ready for crafting.

After harvest, mature loofah fruits grown for sponge production are usually soaked in warm water about 15 min or until the skin slips off of the fibrous interior. Most of the seeds can be removed beforehand by breaking the cap off at the blossom end of the fruit and shaking. With seeds, skin, and excess pulp removed, the sponge is often rinsed in a bleach or hydrogen peroxide solution to whiten the fibres. Then, the sponges are hung to dry.

Another cleaning method, which can produce better quality sponges, starts by harvesting the fruit just as it begins to turn brown. Squeezing the fruit causes the skin to pop loose from the fibres. Then, break off the blossom end cap and pull a strip of rind down the length of the fruit to release the pulpy, fibrous interior. Rinse the sponge clean of excess pulp and seeds, and hang to dry.

Seeds

Cucurbit seeds will continue to develop even after the fruit is removed from the vine. If fruits are harvested before they are fully mature, due to impending frost or other reasons, it is advisable to store them for 1 or 2 months before extracting the seeds.

Whereas seeds of loofah and bottle gourd are extracted from dry fruits and require little or no processing before storage, mature seeds of most cucurbits need to be cleaned of wet, adherent flesh. If the seeds reside in the hollow cavity of the fruit (e.g. pumpkins and some melons), they need only be separated from the placental tissue, washed and dried. Embedded seeds of watermelon, cucumber, melon, wax melon and squash can be removed by chopping and smashing the fruits and adding water to the mixture; seeds will sink and the flesh debris, which floats, is poured off. For commercial seed harvesting, special machinery performs these operations in the field. Recent inventions include the seed sluice for small plots, the bulk seed extractor and the single-fruit seed extractor, which is used in the glasshouse or laboratory to extract cucumber seeds (Wehner and Humphries, 1995).

Removing the persistent placental material encasing cucumber and bitter melon seeds is aided by fermentation. The water and seed mixture is allowed to sit for 1–2 days, depending on the temperature (20–35°C). Fermentation is complete when seeds settle to the bottom of the container and the placental material floats. The seeds are then rinsed and set out to dry.

A dilute solution of hydrochloric acid or ammonia can be used to clean cucumber seeds more quickly. After vigorous stirring, the seeds are free of flesh in 30 min or less. They must be rinsed before drying. This or any other mechanical means of seed cleaning should not be used with bitter melon seeds, which are easily broken.

Cucurbit seeds are typically spread out to dry under warm (<35°C), arid conditions. Commercial producers use forced air warmed by propane heaters and flat drying beds or large rotary dryers. Seeds ready for storage snap instead of bend and have a moisture content of about 5%. Dry seeds should be stored in dark, airtight containers at about 5°C and 25% RH. Silica gel can be added to absorb moisture. Under these conditions, cucurbit seeds may remain viable for as many as 10 years or more.

The chayote seed can only be stored within the fleshy fruit. Mature fruits selected for propagation should be kept in a dark, cool (ca 5°C) area for up to 6 months. Some shrivelling and decay of the fruit may occur during this time. When the next growing season arrives, the entire sprouted fruit is planted.

7

DISEASES AND NEMATODES

INTRODUCTION

Cucurbits suffer from numerous bacterial, viral, mycoplasmal and fungal infections (Blancard *et al.*, 1994; Zitter *et al.*, 1996). Diseases attack cucurbits at every stage of development, from damping off at the beginning of germination to postharvest fruit rots. Reliable productivity and quality of cucurbits is dependent upon adequate pathogen control.

Ultimate disease control is achieved with genetically resistant crops. The first crop ever bred for disease resistance was a cucurbit: a watermelon cultivar ('Conqueror') resistant to fusarium wilt was released by the USDA in 1911. Melon growers also have benefited from selection for genetic resistance; the important 'PMR (Powdery Mildew Resistant) 45' cultivar, introduced over half a century ago, is still grown today. However, the greatest progress in breeding cucurbits for disease resistance has been achieved with cucumber. Sources of resistance have been found for most of the major pathogens of cucumber, and scientists have developed germplasm with multiple disease resistance. For example, 'Wisconsin 2757' is resistant to scab, cucumber mosaic virus (CMV), bacterial wilt, angular leaf spot, anthracnose, downy mildew, powdery mildew, target leaf spot and fusarium wilt. Squash breeding for disease resistance has lagged behind that for cucumber, melon and watermelon, but recently, virus resistant cultivars have been introduced. Research is under way, both by conventional breeding and biotechnology methods, to develop squash cultivars with resistance to multiple pathogens.

Where breeding has not yet produced disease resistance, various preventative and control measures are needed to reduce crop losses. Phytosanitary procedures, including crop rotation and seed treatment, are employed to avoid infection of susceptible plants. Growers often apply chemicals (e.g. fungicides, bactericides, nematicides) preventively as well. Insecticides and other means are used to control the insect vectors carrying contagions, especially viruses. Some susceptible cultivars are grafted to the rootstocks of disease resistant cucurbits (see 'Grafting,' Chapter 5, page 129, for more details).

Early disease detection (before visible signs appear) is possible using enzyme-linked immunosorbent assay (ELISA) kits; presence of the infectious agent in the plant is revealed by testing the sap for pathogen-specific proteins. Some kits can be operated in the field to test for various species of fungi, with results known as quickly as within 10 min.

Disease development is often mediated by environmental conditions. For example, untraviolet-B radiation treatment can increase fungal susceptibility in some cucumber cultivars (Orth *et al.*, 1990); however, UV-B radiation can also harm the fungus and impair disease progression. Fusarium wilt infection of watermelon roots is promoted by low soil pH, low to moderate soil moisture (25% saturation), and the use of NH_4 instead of NO_3 fertilizer. Sudden wilt of melon can be caused by a variety of pathogens in combination with certain environmental conditions, e.g. when cool, rainy weather is followed by hot, sunny weather during the fruiting season. In general, environmentally stressed cucurbits are more susceptible to initial infection and subsequent disease development.

BACTERIAL DISEASES

Bacterial Wilt

Bacterial wilt damages cucurbits in North America and, to a lesser extent, in Europe, Africa and Asia. It is seldom a serious threat to watermelon, but cucumber and melon plants may be killed. The bacterium (*Erwinia tracheiphila*) also infects squash, often resulting in an exudation and decay of the fruit from secondary soft rot organisms.

The first evidence of this malady is a dull green area on the leaf at the site of infection. The vascular system becomes plugged with a slimy substance, and one or two leaves, a branch, or the entire plant may suddenly wilt and die. Diseased plants are sometimes diagnosed by cutting a branch to determine if a sticky exudate will form strands to the knife when it is withdrawn from the cut surface.

Erwinia tracheiphila overwinters in cucumber beetles, and these insects disseminate the bacterium the next season when feeding on cucurbits. Thus, controlling striped and spotted cucumber beetles can curtail disease spread. Cucumber cultivars with the non-bitter allele (*bi*), such as 'County Fair' and 'Marketmore 80', are less preferred by cucumber beetles than other cultivars and, therefore, may escape infection. A single dominant allele in cucumber for resistance to bacterial wilt is known, but it has undesirable associated effects when homozygous.

Bacterial Rind Necrosis

Bacterial rind necrosis is caused by a related species, *Erwinia carnegieana*. Brown necrotic areas occur in the rind of melon and watermelon and sometimes extend into the flesh (Fig. 7.1). Cultivars differ in degree of susceptibility; 'Scarlet Trio' watermelon is resistant.

Soft Rot

Soft rot of cucurbit fruits, especially cucumber and melon, is produced by *Erwinia cartovora* and other organisms. Water-soaked blotches quickly develop into a wet, soft rot and the fruit collapses. This affliction is favoured by hot, humid conditions. Careful handling to avoid fruit injury and chlorinated fruit sprays or dips are preventative measures. There are no resistant cultivars.

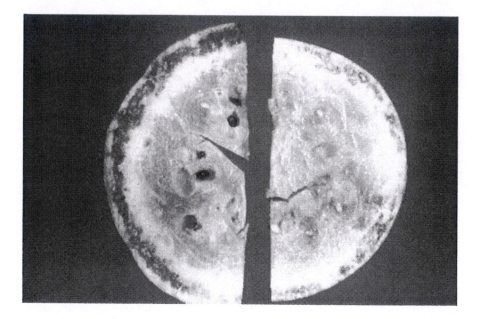

Fig. 7.1. Bacterial rind necrosis of seeded and seedless watermelon, caused by *Erwinia* sp. (T.A. Zitter, photographer, in Zitter *et al.*, 1996, with permission from The American Phytopathological Society.)

Brown Spot

Brown spot of 'Honey Dew' and related melon cultivars results from infection by *Erwinia ananas*. Symptoms are yellow-brown spots up to 4 cm in diameter on the fruit. Damage by other organisms or by mechanical injury predisposes fruits to this disease, which is very serious in a hot, humid environment.

Bacterial Leaf Spot

Bacterial leaf spot of squash plants in India, Australia, Europe, the USA, Japan and elsewhere is caused by *Xanthomonas campestris* pv. *cucurbitae*. Cucumber and other cucurbits are sometimes attacked as well. The bacterium infects leaves through the stomata. Small (2–4 mm) brown lesions, surrounded by a yellow halo, develop on the leaves. Affected areas may coalesce, forming large dead regions delimited by veins. Diseased fruits rot and may be a source of infection for adjoining fruits. Overhead irrigation can spread the bacterium, which survives from one season to the next on plant material in the soil. The pathogen is also seedborne. Genetic resistance has been found in *Cucurbita moschata* and *C. okeechobeensis*.

Angular Leaf Spot

Angular leaf spot is caused by a bacterium (*Pseudomonas syringae* pv. *lachrymans*) of worldwide distribution. *Pseudomonas marginalis* and *P. viridiflava* have also been reported to produce a bacterial wilt of cucumber in Japan. Angular leaf spot is primarily a disease of cucumber (Fig. 7.2), but it damages watermelon, squash and melon as well.

Infection occurs through wounds and stomata. Water-soaked spots develop into small, angular lesions on the foliage, flowers and fruits. The leaf lesions may produce an exudate on the lower leaf surface before dropping out, leaving holes up to 3 cm in diameter. The nearly circular fruit lesions, smaller than those on the leaf, become white to tan and often crack. An infected fruit may rot due to secondary soft rot bacteria entering the fruit through these lesions.

The bacterium overwinters in the soil or on plant debris, and can also reside on the seed surface or below the seed coat. Since it is seed transmitted, the use of disease-free seeds or seed treatment with hot, acidic preparations is recommended. Seeds should be produced where the disease does not occur, with at least a 2-year period between cucurbit crops in a field. The pathogen is also disseminated by the cucurbit leaf beetle (*Aulacophora femoralis*), and control of this insect helps contain angular leaf spot. Keeping the plants dry, by avoiding excessive overhead irrigation in the field or by lowering relative

Fig. 7.2. Lesions on leaves and fruits of cucumber caused by angular leaf spot.

humidity in the glasshouse, reduces bacterial transmission and proliferation. Under wet conditions, farm workers and equipment may spread the disease from infected to healthy plants; therefore, it is best to harvest when the foliage is dry. Fruits should be handled carefully to prevent wounds, e.g. by cutting rather than tearing the fruit from the vine. Copper-containing bactericides have been used to combat this pathogen, but often with only limited success.

For cucumber, this affliction is best avoided by the use of resistant cultivars. 'Cherokee', 'Dasher', 'Poinsett' and 'Premier' are resistant slicing cultivars, and 'Addis', 'Calypso', 'Explorer' and 'Liberty' are resistant picklers. Resistance in cucumber appears to be polygenic.

Bacterial Fruit Blotch

Bacterial fruit blotch has recently become a very serious plague of watermelon. It first became important in 1989 in Florida and other southern states of the USA, and has been a major problem there ever since. The causal agent, *Acidovorax avenae* ssp. *citrulli*, is seedborne and may be transmitted by means other than seed; e.g. it is often mechanically spread in glasshouses where transplants are grown.

Infected seedlings develop dark, water-soaked spots on the foliage and stems, becoming necrotic with yellow margins. Lesions also appear on the fruit, with cracking and exudation making the fruit unmarketable. High temperature and high humidity promote bacterial proliferation.

Seed tests, by growing seedlings in a glasshouse under ideal conditions for disease development, are recommended before planting a field with that seedlot. Testing helps to prevent planting seeds with a high incidence of infection, but cannot be relied on entirely to avert seedborne transmission. Some seed companies use polymerase chain reaction, a DNA amplification process, to test for seedborne bacterial fruit blotch.

Transplants should not be grown in the same glasshouse where there are plants from another seedlot contaminated with the bacterium. Copper-based foliar sprays help control the disease. Avoid planting in fields with recent crops of watermelon or other pathogen hosts, including citron, melon and eggplant. Triploid cultivars are reportedly more resistant to bacterial fruit blotch than diploid cultivars of watermelon.

Salmonella

Salmonella can cause food poisoning. A relatively recent innovation is the sale in supermarkets of melon balls, a convenience food since the consumer does not need to remove the rind and seeds or cut the melon into sections. However, these commercially prepared melon balls sometimes cause food poisoning. The problem is that *Salmonella* bacteria on the external surface of the melon contaminate the flesh when the fruit is cut. Freshly cut melons are quite wholesome and safe, but if stored at the supermarket, restaurant salad bar or elsewhere at warm temperatures, then there is a possibility of bacteria rapidly building up to dangerously high levels

FUNGAL DISEASES

Damping Off

Damping off, a problem of worldwide distribution, can be incited by several pathogens, including *Pythium* spp., *Phytophthora* spp., *Rhizoctonia solani* and other fungi. Damping off often causes poor germination and death of seedlings soon after emergence. It is particularly serious in cool weather, which delays germination and extends the time during which seedlings are most vulnerable to infection. The fungi responsible for damping off of seedlings can also cause a root rot in older plants. Roots of infected cucurbits become water-soaked and flaccid, and the foliage may wilt and die.

Seed treatment with a broad spectrum, thiram-based fungicide reduces damping off, and many seed companies offer treated seeds. Sterilization of plant containers and growing media eliminates some sources of inoculum.

Shallow planting, delaying seeding until warm weather and other procedures that speed germination may lower the incidence of the disease. Cultural practices to keep foliage dry, such as wide spacing and avoiding prolonged irrigation, can be helpful also. Cucurbit seeds should not be sown in poorly drained soils. Rotation, by not planting in fields where cucurbits or other hosts of the pathogens were grown the previous year, should be practised.

Target Leaf Spot

Target leaf spot of cucumber and other cucurbits, which is due to *Corynespora cassiicola*, has occurred for many years in European glasshouses. It is also a problem for field crops in the USA and many other temperate, subtropical and tropical countries.

Target leaf spot has been called the 'leaf fire' disease in reference to the yellow lesions and necrotic spots that develop on foliage of afflicted plants. Angular lesions turn grey and later drop out, giving a shredded appearance and causing defoliation. Young fruits that become infected at the blossom end will shrivel.

The fungus can survive for 2 years on infested plant material, and conidia are airborne. Disease development is favoured by long, hot, humid days, especially when nights are considerably cooler.

Infected vines, a source of inoculum, should be removed from the field, and fungicide application is recommended. Resistance in cucumber is associated with susceptibility to powdery mildew. 'Liberty', 'Astarte', 'Fembasy' and other cucumber cultivars are resistant to target leaf spot.

Alternaria Leaf Blight

Alternaria leaf blight and a storage rot of winter squash, melon, cucumber, watermelon and other cucurbits are caused by *Alternaria cucumerina*, an important pathogen wherever cucurbits are grown. *Alternaria pluriseptata* and *A. alternata* can also cause leaf spotting and fruit rot.

Light brown leaf spots, which are small and round at first, enlarge and cause defoliation. Concentric rings may develop at the lesions on the upper leaf surface. Yield and quality are adversely affected by the reduced photosynthetic area. Spores from diseased leaves infect fruits, especially if the fruits are predisposed by sunscald or chilling injury. Round, sunken, black or brown lesions, measuring about 1 cm in diameter, form and may cause the fruit to rot. Infection typically occurs at the peduncle scar or at a wound.

Crop rotation or the deep ploughing or removal of infected plant debris is recommended, since the fungus is capable of surviving on plant remains for 2 years. Overhead irrigation should be avoided in areas of high rainfall, because long periods of leaf wetness (>8 h) encourage infection. Fungicides are effective, and hot water treatment after harvest can help prevent fruit rot. Melon line MR 1 and 'Edisto 47' are resistant to the disease. Resistant watermelon cultivars include 'Sugar Baby', 'Fairfax' and 'Calhoun Grey'.

Anthracnose

Anthracnose can pose major difficulties for watermelon growers. The disease also attacks cucumber, melon, squash, bottle gourd, fluted pumpkin, pointed gourd, ivy gourd, chayote, snake gourd and other cucurbits. It is caused by *Colletotrichum orbiculare*, a destructive fungus found in India, southeastern USA, and other warm, humid regions throughout the world.

Water-soaked areas on infected leaves turn into brown or black lesions and tan cankers. Conidia from mature lesions are often disseminated by splashing rain. Stem lesions can girdle plants, causing wilting, and the plant may die. Sunken, black, circular or elongated spots occur on fruits, producing a gelatinous mass of pink or salmon-coloured spores when moisture is present. Old lesions become black, and may have white centres with small, black fruiting bodies. Lesions also develop on fruits in storage. The fungus does not require wounds on the fruit for infection.

Disease development is promoted by temperatures of 20–27°C and 100% RH. The pathogen overwinters on infected vines and can be seed transmitted. Sanitation and rotation, disease-free seed, hot water treatment of seeds, fungicides, low temperature treatment of fruits after harvest and resistant cultivars are used to control anthracnose.

'Charleston Grey' has been a leading watermelon cultivar in the USA for many years, mainly because of its resistance to anthracnose. 'Crimsom Sweet' is another important cultivar with anthracnose resistance. The *Ar-1* allele, which is in the genetic background of a number of watermelon cultivars, provides resistance to races 1 and 3, but not to race 2 of anthracnose. Fortunately, race 1 is the predominant race in most watermelon growing areas. 'Au Producer' is resistant to race 2.

Anthracnose resistant cucumber cultivars of the slicing type include 'Poinsett', 'Slicemaster' and 'Dasher II'. Many pickling cultivars, including 'Premier' and 'Explorer', are also resistant.

Cercospora Leaf Spot

Cercospora leaf spot afflicts watermelon, melon, cucumber, fluted pumpkin and other cucurbits in the tropics. The disease develops quickly when the temperature is above 25°C. The fungus (*Cercospora citrullina*) can cause defoliation, due to numerous lesions on the foliage. A good fungicide programme and sanitation, by removing diseased vines, are effective control measures. There are no resistant cultivars.

Powdery Mildew

Powdery mildew is a plague of cucurbits and other crops everywhere. Watermelon is the only important cucurbit crop not subject to this disease in most regions. Two different fungi, *Sphaerotheca fuliginea* and *Erysiphe cichoracearum*, have been reported to cause powdery mildew. It is difficult to distinguish between them because the most diagnostic feature is the cleistothecium, which rarely occurs on cucurbits. Scientists previously thought that *E. cichoracearum* was the causal agent of powdery mildew in most areas, but now *S. fuliginea* is considered the primary agent in Europe and the USA.

First signs of infection are circular white spots, which later enlarge and coalesce until the entire leaf is covered (Fig. 7.3). The white colonies inhabit

Fig. 7.3. White powdery mildew colonies covering a cucumber leaf.

the upper and lower leaf surfaces. The leaves turn yellow, then brown, and die prematurely. Yield is reduced and fruit quality impaired, particularly in melon.

Powdery mildew fungi are unique, in that the mycelia are entirely external. Only the haustoria penetrate the leaves. External mycelia and sporulation make these fungi vulnerable to environmental factors. For example, heavy rains can wash away some spores and mycelia. Consequently, disease development is favoured by hot, dry weather.

Sulphur is an old but effective means of controlling powdery mildew. The dust is applied to cucurbit foliage in the field and is sometimes put on heating pipes of glasshouses to vaporize. At high temperature, sulphur can be phytotoxic; however, some melon cultivars, including 'V-1' and 'SR 91', are able to tolerate sulphur under hot conditions.

Protective fungicides help control powdery mildew when all plant surfaces can be reached. Genetic resistance occurs in *Sphaerotheca fuliginea* to benomyl and triadilmefon, two systemic fungicides used to combat the disease.

Powdery mildew resistance in cucumber is governed by several recessive alleles. Resistant cultivars include Marketmore 76, Polaris 76, Pixie and others. There is a genetic association in cucumber between resistance to powdery mildew and to downy mildew. Although the fungi causing these two diseases are quite distinct, one being an ascomycete and the other a phycomycete, cucumber cultivars resistant to one mildew disease are generally resistant to the other. Also genetically associated with resistance to these diseases is a physiological chlorosis and necrosis that develops in cucumber under certain environmental conditions.

Resistance in melon is race specific. 'PMR 45' has a single dominant allele for resistance. It has been a leading cultivar in western USA and other areas for more than 50 years, but is only resistant to race 1 of powdery mildew. 'PMR 5', 'PMR 6' and some other cultivars are resistant to races 1 and 2 as a result of a single dominant allele and modifiers. Breeding line MR-1 is resistant to all three races.

Powdery mildew resistance has also been found in *Cucurbita* species. Breeders are attempting to transfer this resistance from wild *C. okeechobeensis* ssp. *martinezii* to squash cultivars.

This disease attacks watermelon in some areas of southwestern Asia, although it is not a problem for this crop in most countries. Seshadri (1986) reported that powdery mildew affects watermelon plants in southern India, but not in northern India, presumably due to biotype differences (Fig. 7.4). 'Arka Manik' is resistant to the powdery mildew race infecting other watermelon cultivars in Bangalore.

Another disease, also known as powdery mildew, is caused by *Leveillula taurica*. It attacks cucumber and other cucurbits in the Mediterranean region, and has a wide range of host species in other plant families. This fungus is

Fig. 7.4. Resistant (left) and susceptible (right) watermelon biotypes exposed to powdery mildew.

controlled by the same fungicides used against other powdery mildew pathogens of cucurbits.

Downy Mildew

Downy mildew, produced by the fungus *Pseudoperonospora cubensis*, can be recognized by the checkerboard effect on the leaves, with yellow spots delimited by veins on the upper leaf surface and black sporulation below. This disease, which leads to defoliation and greatly reduced yield, occurs in southeastern USA, Mexico, Israel, Europe, Asia and many other areas worldwide. Although cucumber and melon are particularly affected, squash, watermelon, loofah, pointed gourd, ivy gourd and many other cucurbits are susceptible as well. This pathogen only infects members of the *Cucurbitaceae*.

Downy mildew is favoured by moist conditions. It is an important field disease and can be serious in glasshouses with high humidity. The pathogen overwinters on cucurbits in tropical and subtropical areas; windblown spores from there can later infect plants in more northern areas. It may also persist as oospores on plant debris.

Fungicides are widely employed to combat downy mildew. The use of different chemical preparations in successive applications may prevent increase of a fungal strain resistant to a particular fungicide. Wide plant spacing and other cultural practices that promote rapid drying of foliage can reduce downy mildew incidence and severity also.

The use of resistant cultivars is the best means of control. 'Ashley', 'Stono', 'Palmetto', 'Pixie' and other resistant cultivars bred by C. Barnes have been used as parents to breed many more resistant cucumber cultivars. 'Marketmore 76' is a popular cucumber cultivar that is resistant to downy mildew and other diseases. Resistant melon cultivars include 'Edisto 47', 'Gulfstream', 'Perlita' and others.

Charcoal Rot

Charcoal rot, caused by *Macrophomina phaseolina*, attacks roots, foliage and fruits of many cucurbits. The pathogen inhabits temperate and tropical areas worldwide, and is encouraged by high temperature and soil moisture. Plant parts in contact with the soil develop lesions with black microsclerotia. Cankers, often with concentric rings, occur on the stem and may girdle the plant. Excessive overhead irrigation, and other conditions favouring free moisture on the fruits and foliage, should be avoided. Soil fumigation has met with limited success. There are no resistant cultivars.

Scab

Scab is primarily an affliction of cucumber plants grown in northern areas of the USA, Europe and Asia, where it is cool and moist. The causal agent (*Cladosporium cucumerinum*) also attacks melon, watermelon and squash.

Water-soaked spots form on the leaves of infected plants, later turning white with a chlorotic halo around the lesions. The fungus kills the apical meristem, and diseased seedlings may die. Sunken, dark blotches, sometimes with a sticky exudate, occur on fruits, making them unmarketable. Corky scar tissue develops at the edges of lesions on melon and squash fruits. Melons also become infected at the peduncle scar, in which case a fruit rot ensues.

The fungus overwinters on dead vines of diseased plants, and it can also be seedborne. Scab is controlled by protectant fungicides, but multiple applications may be required, especially during long periods of cool, wet weather. Other control measures include the use of rotation and cultural practices that minimize the presence of free moisture on the foliage.

A single dominant allele for resistance in cucumber has been found and bred into many cultivars, both slicers and picklers. Despite the concern of some for the vulnerability of single gene resistance, scab resistance in cucumber has proved to be stable and durable; scab resistant cultivars have been grown for nearly 50 years, without resistance breaking down due to the occurrence of new fungal races.

Phytophthora Root and Crown Rots

Phytophthora root rot and crown rot, due to *Phytophthora capsici* and other species of *Phytophthora*, cause plants to rapidly wilt and die, often during the fruiting stage. These soilborne fungi infect many cucurbits as well as other crops, causing a foliage blight, with black lesions developing at the crown. Roots of diseased plants have a water-soaked appearance and a black or brown decay. *Phytophthora capsici* also produces a fruit rot of melon, cucumber and watermelon. Soft, sunken spots develop on the fruit and a white mould may form there.

Disease development is promoted by warm weather and wet soil; good drainage helps prevent phytophthora root rot. Damage can be averted by proper cultural practices, such as the use of raised beds and drainage ditches, avoiding overly prolonged irrigation, and not compacting wet soil with heavy machinery. Fungicides are also an effective means of control.

Black Root Rot

Black root rot, incited by *Phomopsis sclerotioides*, occurs in northern Europe, Asia and Canada. It is primarily a malady of glasshouse-grown cucumber and melon, but affects other cucurbits as well. The soilborne fungus invades root hairs, causing a root rot. It can survive in the soil for several years and, thus, is difficult to control by rotation. Sanitation helps prevent dissemination via contaminated soil, and the disease can be avoided by the use of soilless growing media. Systemic fungicides and soil fumigation are effective. Melon and cucumber plants are sometimes grafted to resistant Malabar gourd rootstock.

Root Rot and Vine Decline

Root rot and vine decline of cucurbits result from infection by a number of organisms (Blancard *et al.*, 1994). One of these pathogens, *Monosporasucus cannonballus*, is currently creating a serious disease problem for melon and watermelon plants grown in Texas, Arizona, California, Tunisia, Israel, Japan and Spain. Also affected are cucumber, squash, smooth loofah and bottle gourd. The fungus infects the root system of young plants, killing most of the feeder roots and producing black perithecia on the lateral roots. It causes yellowing and necrosis of crown leaves, often followed by a sudden collapse of the entire plant shortly before maturity. Melon cultivars are being bred for resistance; 'Honey Dew' and 'Deltex' are tolerant.

Brown Rot

Brown rot, also known as choanephora wet rot, affects cucumber, melon, squash, loofah and other cucurbits in the tropics. Spores of the soilborne fungus (*Choanephora cucurbitarum*) are spread by the wind and insects, and cause infection in rainy weather. Blossoms and young fruits become covered by fungal growth, which is white at first and later becomes purplish black. Young summer squashes and other cucurbit fruits may decay. This pathogen is able to survive in the soil for more than a year. Crop rotation, planting on well-drained soils and fungicide application aid in disease control. No genetic resistance is known in cucurbits.

Belly Rot

Belly rot, due to *Rhizoctonia solani*, is a fruit rot of cucumber and melon. It is most serious in southern USA and other areas where warm, moist conditions prevail. Fruits in contact with damp soil may become infected and develop a water-soaked decay, often at that part of the fruit touching the soil. Infection spreads to other fruits during storage. The fungus also causes damping off, stem canker and root rot. It has a wide host range and is frequently present in many soils. Removing infected plants and avoiding excessive irrigation may help prevent disease spread. Pre-planting soil fumigation can be an effective, but expensive means of control. Resistance has been found in USDA plant introductions of cucumber.

Gummy Stem Blight and Black Rot

Gummy stem blight and black rot are caused by a single fungus named *Didymella bryoniae* (sexual stage) and *Phoma cucurbitacearum* (asexual stage). The pathogen produces a brown, blotchy blight with black spots on foliage of watermelon, cucumber, melon, squash, bitter melon, wax gourd, chayote and other cucurbits when conditions are hot and humid. It is an important field disease in southern USA, South America and other subtropical and tropical growing areas, and can be a serious concern for glasshouse cucumber growers. Black rot is also a market and storage disease of cucurbit fruits.

Infection in the seedling stage is often lethal. Tan to black lesions form on the foliage and fruits of older plants. The disease may develop so suddenly that foliage appears to be scorched. The stem can become infected, girdling and killing the plant. There may be gummy, black fruiting bodies on foliage and fruits of some cucurbits.

Infected *Cucurbita pepo* and *C. maxima* fruits typically turn black and quickly rot, but 'Butternut' (*C. moschata*) fruits develop circular brown rings without a soft rot. 'Butternut' squashes often display fruit symptoms in the field, but those of *C. maxima* usually do not succumb to black rot until after harvest. Fruits may become infected through wounds, and careful handling during harvest and shipping can minimize losses. Curing harvested fruits (see Chapter 6 for more details) and low postharvest temperature (but >10°C) reduce the incidence of black rot during storage.

Infected seeds spread the fungus; consequently, only disease-free seeds should be planted. Removing afflicted plants and crop rotation for two or more years are advised. Ventilation in the glasshouse and other measures to reduce humidity are helpful. Irrigation methods that do not leave moisture on leaf surfaces are preferred to avoid infection. Fungicides should be applied early in the day so that the plants do not remain wet overnight. Fungal resistance to benzimidazole fungicides has evolved in some glasshouse growing areas in Europe and in cucurbit fields in eastern USA.

Limited genetic resistance has been found in cucumber, melon, squash, watermelon and wild species of *Cucurbita* and *Cucumis*. 'Congo' watermelon, 'Nicklow Delight' squash (*C. moschata*) and 'Gulfcoast' melon possess resistance alleles.

Fusarium Wilt

Fusarium wilt, due to *Fusarium oxysporum*, is an important blight of melon and watermelon, especially in the Americas, Europe and Asia. It also attacks cucumber. The soilborne fungus penetrates the roots and develops in the vascular system, interfering with water movement and causing a severe wilt. It flourishes when soil temperature is high. Watermelon is particularly affected, and diseased seedlings may die.

Preventative measures include rotation with crops that are not hosts of the pathogen and not planting susceptible cucurbits in fields with a known history of fusarium wilt. Because the fungus exists in the soil for many years, short rotations are inadequate. In Africa, watermelons are rotated with crops such as corn, cotton and legumes every 4–10 years.

Raising soil pH to 6.5 inhibits disease development. In Japan, fusarium wilt has been controlled by grafting melon plants on to Malabar gourd or wax gourd rootstock. Although hot-water seed treatment reduces seed transmission, the use of disease-free seeds, produced where the pathogen does not occur, is recommended.

Different biotypes of the fungus cause the fusarium wilts of different cucurbits. Fusarium wilt of watermelon is caused by *Fusarium oxysporum* f. sp.

niveum, fusarium wilt of melon is due to *F. oxysporum* f. sp. *melonis*, and that of cucumber is caused by *F. oxysporum* f. sp. *cucumerinum*. Three races of watermelon fusarium wilt have been identified. Resistant watermelon cultivars have been available since the first decade of this century; 'Charleston Grey', 'Summit' and other cultivars are resistant to race 1. Four races of fusarium wilt of melon are known. Melon cultivars resistant to races 0 and 2 include 'Iroquois', 'Delicious 51', 'Harvest Queen', 'Honey Dew' and others; 'Charmel' is resistant to race 1. Distinct races of fusarium wilt of cucumber have also been discovered. 'Calypso', 'National Pickling' and other cultivars are resistant to some of these races.

Fusarium Fruit Rot

Fusarium fruit rot, which can be caused by *Fusarium solani* f. sp. *cucurbitae*, is a cosmopolitan and serious affliction of cucurbits, particularly squash. Two biotypes of this fungus are known, one attacking fruits, roots and stems, and the other confined to fruits of the host plant. The pathogen invades the outer part of the stem, causing wilting, brown decay at the base of the stem and collapse of the plant. The stem may become covered with the white or pink fungal growth. Infected fruits have water-soaked areas that develop into a brown or grey dry rot. Other *Fusarium* species, including *F. equiseti*, cause cucurbit fruit rots in which infected areas on stored fruits turn pink, red or purple, with white or pink mycelia.

Because spores are carried externally on cucurbit seeds, seed treatment is used to reduce disease spread. The fungus is persistent in the soil, and a rotation of at least 3 years is recommended. Hot-water treatment (1 min at 57°C) and fungicide application limit pathogen transmission among harvested fruits. There are no resistant cultivars.

Verticillium Wilt

Verticillium wilt, due to *Verticillium dahliae* or *V. albo-atrum*, attacks melon, squash, cucumber and other cucurbits worldwide. It is a soilborne field disease which also damages glasshouse cucumbers. Symptoms are similar to those of fusarium wilt, but verticillium wilt can be serious in areas with lower soil temperature. Wilting often occurs near plant maturity, and the chlorosis and necrosis of crown leaves may be more severe than with fusarium wilt.

Verticillium dahliae causes cavity rot of wax gourd fruits. Stored fruits may appear outwardly sound, but internally, they contain decay and mycelia. The fungus also invades wax gourd roots, stems and peduncles.

Verticillium wilt fungi can persist in the soil for years, so cucurbits should not be replanted in a field with a history of the disease. Melon cultivars differ in their susceptibity to verticillium wilt; 'Persian' and 'Casaba' are more easily infected than 'Honey Dew' and 'PMR 45'.

Cottony Leak

Cottony leak, caused by *Pythium aphanidermatum* and other species of *Pythium*, is a fruit rot of cucumber, squash, melon and watermelon. Cottony fungal growth develops from water-soaked lesions at infection sites and spreads over the fruit, which may produce a watery discharge. These pathogens also cause damping off of seedlings and root rot.

Pythium species have a very wide host range and persist in soil for years, making rotation largely ineffective for control. Moist conditions and warm temperature in the field or during storage promote disease development. Plastic mulch should be avoided in rainy areas, because the depression where the fruit sits will collect water. However, other artificial barriers such as wood blocks can be placed between the fruit and soil to avoid infection.

Sudden Wilt

Sudden wilt of melon is an abrupt collapse of the plant, often just before harvest, when plants are stressed by a heavy fruit load. It frequently occurs following cold nights. Various organisms, which might not be the cause, have been isolated from affected plants, but *Pythium ultimatum* and related species and CMV have been implicated as causal agents.

Stem End Rot

Stem end rot, a product of *Lasiodiplodia theobromae*, is a common storage disease of watermelon in the Americas. It also afflicts melon, squash, cucumber and other vine crops.

Although the disease is primarily a postharvest problem, causing rot of fruits in shipment, in storage or at market, the fungus also proliferates in the field. It causes wilting and a foliage blight, producing lesions with a gummy exudate on stems and leaves. Infected fruit tissue becomes soft and water-soaked; later, the watermelon shrivels, turning brown at the peduncle end. Dark grey mycelia and black fruiting bodies form in the lesions and secondary organisms may grow there as well. Diseased fruits sometimes have an odour.

Careful handling to prevent fruit injury helps to avert infection. Watermelons should be cut, not pulled, from the vine at harvest, and a long peduncle should be left on the fruit. For shipping, the peduncle can be recut and treated with a copper-containing paste. Low temperature retards development of the fungus, and appropriate postharvest conditions reduce losses from stem end rot. Application of fungicide is important for control in the field.

Cottony Soft Rot

Cottony soft rot, due to *Sclerotinia sclerotiorum*, produces stem cankers and fruit rots on many cucurbit crops worldwide. The first sign of this disease is usually white, cottony fungal growth on the stem. Infected plants may develop a white mould with black sclerotia inside the stem and then die. The fungus also grows on leaves and fruits. Fruits are often infected through a withered flower still attached to the developing ovary. A soft rot and white, cottony mould with embedded black sclerotia quickly envelops the fruit. The disease spreads by contact from one fruit to another in storage.

The fungus also causes white mould of beans, cabbage and other plants. Buildup of this contagion in a field planted with one of these crops may lead to considerable destruction of squash or melon plants grown in that field the following year. Since there are other pathogen hosts and sclerotia survive in the soil for years, rotation away from cucurbit crops may not provide adequate control. Sanitation, fungicide application, care in irrigating and harvesting, good storage conditions (10–15°C and 50–75% RH) and the prompt removal of infected fruits from storage are recommended control measures.

Southern Blight

Southern blight is caused by a related soilborne fungus (*Sclerotium rolfsii*), which produces a fruit rot of melon, squash and watermelon, especially in tropical and subtropical areas. Infected plants often wilt and turn yellow. They may die due to girdling of the stem at its base. White fungal growth with embedded brown sclerotia develops on the stem and on fruits in contact with the ground. Rotation, sanitation, raising soil pH to 7.0 and fungicides aid in the containment of this disease. Resistant cultivars are not available.

Grey Mould

Grey mould, due to *Botrytis cineraria*, is a problem for cucumber plants grown in cool, cloudy weather. It attacks glasshouse cucumbers everywhere, and also afflicts melon and squash. Infection often occurs through dead tissue, such as senescent petals and other flower parts. Abundant mycelia and spores develop on the blossom end of the fruit, causing it to rot. Brown lesions form on the foliage and are covered by a mass of grey spores. Good ventilation, warm temperature and avoiding free moisture on the leaf surface are preventative measures in the glasshouse.

Septoria Leaf Spot

Septoria leaf spot is a leaf blight and fruit spot of melon and squash caused by *Septoria cucurbitacearum*. The pathogen thrives in temperate areas with rainy weather and temperatures of 16–19°C. Small, white spots with brown halos appear on leaves of infected plants. White spots may also occur on the fruits, making them unmarketable. Pycnidia, which are small, black fruiting bodies, develop in the lesions. The fungus overwinters on infected plant debris; hence, rotation and ploughing after harvesting an infected field can help prevent persistence of the disease. The pathogen is generally controlled by chemical means, even though some fungal strains have developed resistance to some fungicides.

Ulocladium Leaf Spot

Ulocladium leaf spot, due to *Ulocladium cucurbitae*, produces a leaf blight on field-grown cucumber plants in New York and on field- and glasshouse-grown cucumbers in the UK. Irregular tan lesions with a buff centre, measuring 0.3–3.0 cm in diameter, develop on leaves and coalesce to cause defoliation. This minor disease can be controlled by fungicides, rotation or the use of resistant cultivars such as 'Dryden' and 'Albion'. *Ulocladium atrum* also occurs on cucumber in the UK and France, but is usually of little importance.

Rhizopus Soft Rot

Rhizopus soft rot, caused by *Rhizopus stolonifer* and related species, produces a postharvest fruit rot of melon, squash, cucumber, watermelon and other cucurbits. The disease is promoted by the presence of free water and by fruit cracks or wounds, which may become sites of infection. Infected tissue becomes soft and water-soaked. Careful handling and the maintenance of low temperature during shipping and storage discourage disease development. These fungi have a very wide host range and survive in the soil from one year to the next.

VIRAL AND MYCOPLASMAL DISEASES

Over 30 viruses infect cucurbits, nine of which are seedborne (Zitter *et al.*, 1996). Viruses can be a very serious problem because there are no chemical protectants against them. The best control measure is a resistant cultivar, if available. Eliminating weeds, which are often the source of the disease, and controlling aphids and other insects that transmit the virus can also be beneficial. However, aphids may carry a virus to cucurbits from far away, so weed control alone in a cucurbit field may not contain the disease. There has been some success with reflective mulch to repel aphids, and rowcovers can exclude insect vectors altogether. After a virus is introduced into a field, it is often spread by mechanical means. When there are multiple harvests, such as with cucumber and summer squash, pickers can be a major cause of disease spread.

Cucumber Mosaic Virus

Cucumber mosaic virus (CMV) stunts seedling growth (Fig. 7.5), causes mottling and distorted shape of leaves and fruits, and reduces yield. Cucumber, squash, melon and many other cucurbits are susceptible, but watermelon is less affected by most strains of CMV. The disease is of worldwide distribution and importance.

CMV is spread by aphids from its very wide range of host plants. Cucumber beetles also carry the disease. Insecticides used to kill the vectors are often ineffective for control of CMV, since the insects may disseminate the virus before being killed. The virus is easily transmitted mechanically, and is often spread by workers harvesting fruits.

CMV resistance seems to be common among wild *Cucurbita* species. Interestingly, all wild taxa tested by Provvidenti *et al.* (1978) were resistant to this important virus except *C. pepo* var. *texana* and *C. maxima* ssp. *andreana*, the closest wild relatives of two of the domesticated squash species. Resistance has been found in some landraces of *C. maxima* and *C. moschata* from South America and Nigeria, respectively.

Most modern cucumber cultivars are resistant to CMV. In contrast, all important melon cultivars are susceptible, but resistance occurs in some members of the Conomon Group. In recent years, genetic engineering techniques have been employed to transfer CMV coat protein-derived resistance to cultivars of squash, cucumber and melon; see Chapter 3 for more details.

Cucumber Green Mottle Mosaic Virus

Cucumber green mottle mosaic virus (CGMMV) attacks cucumber, melon and watermelon in subtropical areas, including India. It is also a serious afflication of glasshouse cucumbers in Europe, Japan and elsewhere. Infection occurs through the roots, coming from diseased plant debris in the soil. Leaf

Fig. 7.5. Squash (*Cucurbita pepo*) seedlings inoculated with cucumber mosaic virus. The left seedling was resistant; the right was susceptible and shows symptoms.

distortion, mottling and vein clearing occur, and yield may be reduced. Symptoms vary according to the strain of the virus.

The virus is seed transmitted; hence, the use of disease-free seeds is important. Seed treatment with high temperature, Na_3PO_4, or another appropriate chemical is recommended. Sanitation is vital, since the pathogen can be transmitted from roots of previous crops and is also spread mechanically. Resistance to CGMMV has been found in melon.

Squash Mosaic Virus

Squash mosaic virus (SqMV), which has a worldwide distribution, is a very destructive disease of squash and melon. The virus causes severe leaf distortion, often resembling injury from auxin-type herbicides. There may be enation on the lower leaf surface. Fruits are frequently mottled and deformed, and yield reduced.

Insects, including cucumber beetles and *Epilachna chrysomelina*, are contagion vectors. SqMV overwinters in cucumber beetles and infected seeds. To prevent seedborne infection, plant only virus-free seeds and do not save seeds from afflicted plants. Insect control and removal of diseased plants reduces spread of the virus. Weed control is helpful, since *Chenopodium* species can become infected with SqMV and transmit the pathogen by seed. If the virus is present in a field, cultural implements should be disinfected to reduce mechanical transmission; otherwise, the virus can be spread by harvesting and other cultural operations. Immunity or high-level resistance to SqMV has not been found in *Cucurbita*, but Provvidenti *et al.* (1978) reported that *C. ecuadorensis* recovered from symptoms after infection. Resistance exists in bottle gourd.

Watermelon Mosaic Virus

Watermelon mosaic virus (WMV), previously known as WMV-2, plagues squash, melon and other cucurbits in Europe and North America. It also attacks plants of other families. The virus causes stunting and mottling of the foliage, and unsightly green rings may develop on squashes normally having a yellow rind (Fig. 7.6).

WMV is transmitted mechanically and by aphids. Weed control reduces sources of inoculum. Application of insecticide or mineral oil curtails transmission by aphids. Aluminium mulch, rowcovers and other measures that discourage insect visitation are also used. 'Sweet Slice' cucumber,

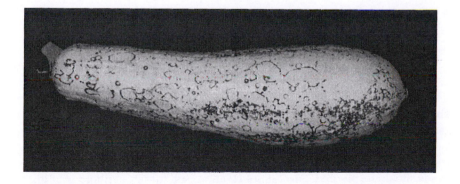

Fig. 7.6. The green rings on this yellow summer squash fruit were caused by watermelon mosaic virus. (Whitaker and Robinson, 1986. Reprinted courtesy of Chapman & Hall, New York.)

'Freedom II' summer squash, the winter squash 'Redlands Trailblazer' (*C. maxima*), melon line B66-5, 'Cow Leg' bottle gourd and the watermelon relative, colocynth, are resistant to WMV.

Papaya Ringspot Virus

Papaya ringspot virus (PRSV), formerly known as watermelon mosaic virus-1, is a serious disease of many cucurbits in warm regions worldwide. It causes leaf mottling and distortion, and infected fruits may become bumpy and deformed. The virus overwinters on cucurbits in tropical and subtropical areas, and can be dispersed from there to other areas by aphids. Sanitation and cultural procedures to control aphids restrict dissemination of the disease. PRSV is often spread mechanically by pickers. 'Sweet Slice', 'Douceur' and 'Zena' cucumbers, 'Redlands Trailblazer' squash and 'Cow Leg' bottle gourd are resistant.

Tobacco Ringspot Virus

Tobacco ringspot virus (TRSV) is transmitted to squash, cucumber, watermelon and other cucurbits by nematodes. It can also be spread mechanically and by insects. Tiny yellow-brown spots with a yellow margin occur on leaves. Small water-soaked spots may develop on fruits. The virus has a very wide range, overwintering on many weed species of different plant families. Weed control can help keep the virus in check. Soil fumigation with nematicides curtails transmission by nematodes, but at an expensive cost. Resistance has been found in melon, squash and bottle gourd.

Melon Necrotic Spot Virus

Melon necrotic spot virus (MNSV) produces necrotic lesions on the foliage of melon and cucumber in Mediterranean countries, Japan and elsewhere, especially on crops grown in glasshouses or other protected environments. The disease is transmitted by a soil fungus, *Olpidium radicale*, which enables the virus to enter the plant through the roots. This fungus also transmits cucumber necrosis virus and tobacco necrosis virus to cucumber. Cucumber beetles reportedly are another vector of MNSV. The virus is readily spread mechanically and can be seedborne as well. Consequently, only disease-free seeds should be planted. Sanitation, rotation, soil fumigation and grafting to a resistant rootstock of Malabar gourd are preventative measures.

Clover Yellow Vein Virus

Clover yellow vein virus (CYVV) was formerly known as the severe strain of bean yellow mosaic virus. It produces small, quite noticeable spots on leaves of infected squash plants. CYVV can cause economic losses to cucumber, but is usually not as destructive to squash.

Zucchini Yellow Mosaic Virus

Zucchini yellow mosaic virus (ZYMV) is a serious threat to cucurbits throughout the world today, although it was not recognized before 1980. The virus is very damaging to squash, melon and watermelon. Though generally less destructive to cucumber, ZYMV can still cause economic loss to this crop. It also attacks many other cucurbits.

ZYMV causes leaf mottling and extreme stunting. Diseased fruits are often very distorted in shape and unmarketable. Squash cultivars normally having yellow fruits may have partially or entirely green fruits when infected. ZYMV is easily transmitted mechanically and by aphids.

Different strains and pathotypes of ZYMV exist. A single allele has been found in melon for resistance to one pathotype, but not to another. The *zym* allele discovered in a cucumber from Taiwan provides resistance to both of these pathotypes. 'Sweet Slice' and 'Taichung Mou Gua' cucumber cultivars are resistant. Resistant squash cultivars include 'Jaguar', 'Tigress' and the transgenic 'Freedom II' (*C. pepo*); 'Nigerian Local' and 'Nicklow Delight' (*C. moschata*); and 'Redlands Trailblazer' (*C. maxima*). Resistance also occurs in bottle gourd.

Squash Leaf Curl Virus

Squash leaf curl virus (SLCV) has become a problem in recent years for squash growers in southwestern USA and Mexico. Related viruses attack watermelon and melon in the Middle East and elsewhere. SLCV causes a severe leaf curl (with margins curling upward), mottling and enation. Infected fruits may be small, deformed and unmarketable.

The virus, which is prevalent in subtropical areas where cucurbits are continuously cropped, is transmitted by the sweet potato whitefly. Avoiding growing cucurbits and other hosts of this insect for a period of time in an area can keep the disease in check by curtailing buildup of the vector. There are no resistant cultivars, although tolerance has been reported in *C. pepo*, *C. moschata* and wild species of *Cucurbita* and *Cucumis*.

Muskmelon Yellows Virus

Muskmelon yellows virus (MYV) infects melon and cucumber plants in Spain, The Netherlands, Japan and other countries. It has become especially troublesome in recent years on melon plants grown in plastic tunnels and other protected environments. MYV causes chlorosis and reduces yield and quality. Similar symptoms are produced on cucumber and squash in The Netherlands and Japan by pathogens considered to be other strains of MYV.

The virus overwinters on a wide range of weed hosts, and is carried from these hosts to cucurbits by the glasshouse whitefly. Chemical treatments to control whiteflies may restrict spread of the disease, but seldom can a large whitefly population be eradicated and virus transmission eliminated. Disease spread is reduced by sanitation, weed control and rowcovers that exclude insects. Tolerance has been found in Asiatic melon landraces. African horned cucumber and teasel gourd are resistant.

Lettuce Infectious Yellows Virus

Lettuce infectious yellows virus (LIYV) produces symptoms similar to those of MYV on melon and other cucurbits. LIYV, which has been a problem in California, Mexico and the Middle East, also infects lettuce and other species. Control of the sweet potato whitefly, which carries LIYV, is the most practical means of preventing disease transmission. Melon cultigens differ in susceptibility to LIYV, but no cultivar has been found to be highly resistant.

Aster Yellows

Aster yellows, formerly thought to be caused by a virus, is a phytoplasmal disease. The pathogen overwinters on infected weeds and is spread to annual cucurbit crops by leafhoppers. It causes chlorosis and stunting of squash and cucumber plants. Leaves, flowers and fruits become malformed. Weed and insect control are somewhat effective.

NEMATODES

Root knot nematodes (*Meloidogyne incognita* and other *Meloidogyne* species) invade root tissue, causing galling of cucurbit roots. Adult nematodes are embedded in the root galls. Infected plants are stunted and wilted, and yield is reduced. Nematodes also damage cucurbits by transmitting TRSV.

Root knot nematodes cause substantial losses of melon, watermelon, squash, cucumber and many other cucurbits. They are usually most prevalent on sandy soils, and are seldom serious in the fields of northern areas, where severe winter temperatures inhibit their overwintering. However, they may infest soil beds in glasshouses anywhere.

Control measures include soil fumigation, rotation with a crop not included in the wide host range of the nematodes affecting cucurbits, and grafting to a resistant rootstock. Cucurbits that are resistant include African horned cucumber, bur gherkin and wax gourd. African horned cucumber has been used as a resistant rootstock, but attempts to transfer its resistance to melon and cucumber have been unsuccessful. Resistance to some species of *Meloidogyne* was found in *Cucumis sativus* var. *hardwickii*, but this taxon was susceptible to both races of *M. incognita* tested (Walters *et al.*, 1993). Cucumber, watermelon and melon cultivars differ in their susceptibility to species of *Meliodogyne*, but none is resistant.

A recent experiment with cucumber plants grown on nematode-infested soil (Hanna *et al.*, 1993) tested the effects of double-cropping with a nematode resistant cultivar of tomato, using cucumber transplants instead of direct seeding, and pre-plant field application of the nematicide ethoprop. Preceding cucumbers with a nematode resistant tomato significantly reduced nematode populations in the field. Although using cucumber transplants and applying ethoprop did not reduce overall nematode population size, cucumber yield and growth were improved.

8

INSECTS AND SPIDER MITES

Insects are often a major obstacle to the successful production of cucurbits (York, 1992). They can severely reduce yield, injure or kill plants, spread disease and adversely affect fruit quality. Spider mites can also be a problem, particularly with glasshouse cucumbers.

Integrated pest management requires an understanding of the biology and life history of the living organisms that interact with cucurbits. Management begins with cultural control methods, such as crop rotation, deep ploughing of crop residues, control of alternate hosts (i.e. weeds) and the use of trap crops. Because squash plants are strong attractors of the sweet potato whitefly, they can be planted as a trap crop to protect tomatoes (Coleman, 1995). Rowcovers are effective at restricting pest (e.g. cucumber beetle) access to young cucurbit crops, and UV-reflective aluminium plastic mulch discourages visitation by virus-carrying aphids. In the glasshouse, sticky boards and similar traps catch whiteflies and other flying insects.

Cultural and physical controls alone are rarely sufficient for cucurbit pest management. Foliar application of chemicals is the most common means of controlling insects and mites in the field, and aerial fumigation is also used in the glasshouse. Soil fumigation, although expensive, can eradicate soilborne pests.

The trend in recent years has been to reduce the amount and spectrum of chemical pesticides applied to cucurbits. Instead of routinely applying insecticides on a periodic basis, as was often done previously, growers or consultants make inspections and count insects to determine if insecticide application is justified. In addition, 'biorational' pesticides have been developed which have less impact on the overall environment and non-target organisms. For example, insecticidal soaps and oils are less toxic chemicals for whitefly control than broad-spectrum preparations containing pyrethroids and organophosphates, although the latter generally are more effective at reducing pest populations. Biorational insecticides also include microbial-based preparations and pheromones that modify insect behaviour. The bacterium *Bacillus thuringiensis* attacks squash vine borers and other cater-pillars, whereas the fungus *Verticillium lecanii* is used in European glasshouses to infect whiteflies, thrips and aphids. Beneficial, insect-attacking nematodes (e.g. *Steinernema carpocapsae*), applied to the soil or plant through watering or spraying, injure various cucurbit pests, including squash vine borers, cucumber beetles, cutworms and fruit fly larvae.

Beneficial insects are also employed for biological management. For example, the parasitic wasp *Encarsia formosa* can keep glasshouse whitefly populations in check. For cucumber, the recommended treatment is to release one parasite for every two plants while whitefly numbers are still low. A predator of spider mites, *Mesoseiulus longipes*, also works well in warm glasshouse environments. Aphid control for indoor and outdoor cucurbit crops can be achieved with *Aphidoletes aphidimyza* and various beneficial ladybird beetles, respectively. The concurrent use of chemical insecticides with biological controls needs to be carefully evaluated and monitored so that the beneficial insects and pollinators are not killed along with the pests.

Breeding cucurbits for insect and mite resistance has great potential, but has not progressed nearly as far as breeding for disease resistance. Very few cucurbit cultivars have been bred for resistance to any insect, although sources of resistance are known (Robinson, 1992). Some landraces developed hundreds of years ago are resistant, even though they were not intentionally bred for pest resistance.

An important problem in breeding cucurbits for insect resistance is that an allele for resistance to one insect may render the plant more susceptible to other insects. This is particularly true for genes influencing cucurbitacin content. These very bitter and highly toxic compounds evidently evolved as a defence mechanism to protect cucurbits from insects and other herbivores. However, during the course of evolution, insects of the *Chrysomelidae* family developed mechanisms to tolerate cucurbitacins (Metcalf and Rhodes, 1990). Today, cucumber beetles and related insects, such as the red pumpkin beetle and corn rootworms, are not repelled by cucurbitacins like many other insects are, but instead are attracted to these compounds. Consequently, cucurbit plants lacking cucurbitacins are typically resistant to cucumber beetles but more susceptible than normal to spider mites, aphids, thrips and *Margaronia hylinata*.

Cucumber Beetles

Cucumber beetles can cause very severe damage, and are the most important insect pest of cucurbits in many areas of the western hemisphere. *Acalymma vittatum* (Fig. 8.1) and several species of *Diabrotica* attack young seedlings, riddling the cotyledons and leaves with holes, damaging the roots, spreading

Fig. 8.1. Adult striped cucumber beetle, *Acalymma vittatum*. (Courtesy M.P. Hoffmann, photographer.)

diseases and even killing some plants. Foliar damage reduces fruit set and lowers yield.

Cucumber beetles, which often appear in very large numbers when cucurbit plants are still in the young seedling stage, attack during all stages of plant development. They may feed on blossoms, especially those of *Cucurbita maxima*, as well as on roots, fruits and foliage.

In addition to the extensive damage they cause by feeding, cucumber beetles are a major vector for bacterial wilt. The bacterium causing this disease overwinters in the intestines of cucumber beetles, and the beetles transfer the disease to cucurbits the following season. Cucumber beetles also transmit squash mosaic virus (SqMV) and other viruses.

The adult beetles are about 0.5 cm long and yellow, with black stripes or spots. They overwinter as unmated adults, under debris in fields or woodlands, emerging from hibernation in the spring when the temperature rises above 10°C. After mating, the female lays orange-yellow eggs around the base of cucurbits or other host plants. The larvae, which are white with black ends, hatch and burrow into the soil, where they feed on roots and underground stems. The white pupae persist in the soil for about a week; then, adults migrate above ground where they feed on foliage. There may be two generations per year in warm climates and one in cooler areas.

Several *Cucurbita pepo* cultivars, including 'Yellow Crookneck' and 'Acorn', have lower cucurbitacin content in their cotyledons and are less preferred by cucumber beetles than cultivars such as 'Zucchini' and 'Caserta', which have more cucurbitacin. This difference in cucurbitacin content and attractiveness to cucumber beetles is conditioned primarily by a single gene (see 'Squash genes,' Chapter 3, page 45). Genetically controlled cucumber beetle resistance has also been reported in *C. moschata*, melon, watermelon and cucumber.

Cucurbitacins can be used as bait for cucumber beetles. Bitter fruits of *Cucurbita maxima* ssp. *andreana* or another cucurbit are cut open and laced with an insecticide. Numerous cucumber beetles are lured to each cut fruit by its high cucurbitacin content and perish after eating the deadly contents. The commercialization of poison baits containing cucurbitacin extracts is being investigated.

Some insecticides control cucumber beetles. Autumn ploughing and other field sanitation measures may reduce pest population size in a field the following year, but sanitation and rotation may be inadequate control measures alone since the adult beetles can immigrate from other fields.

Red Pumpkin Beetle

Red pumpkin beetle (*Aulacophora foveicollis*) is closely related to the cucumber beetle. It is one of the most significant cucurbit pests in India, and also attacks plants in other Asian countries, the Middle East and southern Europe. This insect damages melon, loofah, bottle gourd, watermelon, squash and other cucurbits, particularly in the seedling to young plant stage. Adults feed on cotyledons, immature leaves and flowers.

The adult red pumpkin beetle is approximately 7 mm long and similar in appearance to the spotted cucumber beetle, except for its red-orange colour. The adult overwinters in the soil. In the spring, eggs are laid on the soil under debris. Later in the season oviposition may also be on the rinds of fruits touching the ground. Larvae pupate in the soil. Two or more generations occur per year.

The red pumpkin beetle is controlled with insecticides. Resistance has been found in melon, watermelon, bottle gourd, loofah and species of *Cucurbita*.

Epilachna Beetles

Epilachna beetles (*Epilachna* spp.) are primarily pests of cucurbits and other crops in Africa. Watermelon, cucumber, summer squash and ivy gourd are affected. The adult beetles, measuring 6–8 mm long, are reddish to brownish-yellow with black spots. Clusters of pale yellow eggs are laid on the underside of the leaf. The larvae and adults feed on the leaf, sometimes stripping the blade to the midrib. These insects also gnaw holes into stems and fruits. They are controlled with insecticidal sprays.

Squash Vine Borer

Squash vine borer (*Melittia cucurbitae*) prefers squash, especially cultivars of *Cucurbita pepo*. Cucumber, watermelon, melon and other cucurbits can also be damaged, but the insect may not complete its life cycle on these hosts. Although the squash vine borer is restricted to the western hemisphere, from South America to Canada, the genus *Melittia* is of worldwide distribution, and other species can attack squash species that are resistant to *M. cucurbitae*.

The orange and black moth, about 1.5 cm long, has opaque forewings and transparent rear wings with a wingspan of about 35 mm. The wasp-like moth lays small, reddish brown eggs singly in the daytime at the base of squash plants, mostly on the stem or leaf petioles. The larvae hatch and tunnel into the stem of the plant. This often causes the stem to collapse, and the plant wilts and may die. The frass around the hole where the borer entered the stem is indicative of the insect's presence. The larvae, which are white with black head and thoracic shield when hatched, grow to 25 mm from an initial length of approximately 2 mm. After feeding within the stem for 2–4 weeks, the larvae fall to the ground and overwinter and pupate in the soil. Pupae are brown and about 16 mm long. There is one generation per year in temperate areas; additional generations may develop in regions with a longer growing season.

Home gardeners sometimes crush the eggs, mound soil about the stems of plants wounded by a borer in order to promote adventitious rooting, or kill borers with a sharp instrument. Commercial growers more often rely on insecticides, which are best applied to the base of plants early in the season. Control by insecticide may be difficult after eggs have hatched, since the borers are embedded in the stem where they are protected from externally applied chemicals. Rotation is recommended to avoid planting in fields with soilborne pupae.

Cucurbita moschata and *C. argyrosperma* are resistant. It has been suggested that the squash vine borer resistance of *C. moschata* is associated with its hard, woody stem. Genetic differences in volatile chemicals of the host plant are also considered to influence oviposition.

Melon Aphid

Melon aphid (*Aphis gossypii*) attacks not only melon, but also cucumber, squash and other cucurbits. In large numbers, these insects reduce yield and quality by sucking the plant juices. Curling of young leaves and blackened foliage or fruits, due to fungal growth stimulated by excreta of aphids, are signs of heavy aphid infestation. Frequently, even more serious losses are caused by melon aphids transmitting viral diseases; e.g. CMV, ZYMV, PRSV and WMV.

Aphids are small, sucking insects which often feed on the lower leaf surface or stem tips. They may be green, yellow or black; melon aphids are usually dark green. Both winged (alate) and wingless (apterous) forms occur for the same species. Winged forms can fly or be carried by the wind over long distances. They overwinter as eggs in temperate regions; in the tropics, all stages of the life cycle may exist throughout the year. Aphids have a very wide host range, which includes a number of different plant families. They can reproduce sexually or by parthenogenesis, quickly building up to large populations under favourable conditions. Often, numerous generations occur per year.

Some aphicides are systemics applied to the foliage or soil. Contact insecticides should target the undersides of leaves. Fungal preparations of *Verticillium lecanii* are effective in glasshouses when conditions are warm and humid.

Good weed control also helps limit aphid populations. Early and late plantings of cucurbits should be geographically isolated, if possible, to minimize the spread of insects from one field to another. Brightly coloured aluminium mulch, which repels aphids, is most effective early in the season, before it becomes covered by cucurbit vines.

A genetic system to simulate the effect of aluminium mulch has been proposed as a means of deterring aphids and preventing aphid-transmitted viruses. Oved Shifriss has developed a summer squash line (NJ 260) with this characteristic; it has 'silvery' leaves, due to gene *M* and modifiers.

Insect resistance is classified into three categories: non-preference (less attractiveness to insects), antibiosis (adverse effect on the insect feeding on that host), and tolerance (recovery from insect feeding). Each of these three factors has been determined to be involved in the aphid resistance of melon PI 371795 (Bohn *et al.*, 1972). This resistance is due to a single dominant allele, *Ag*. USDA researchers have backcrossed this allele into 'AR Hales Best Jumbo', 'AR Topmark' and 'AR 5'. Another allele (*Vat*), derived from Spanish melon landraces, reduces the amount of the virus transmitted by melon aphids. Genetic variation occurs in aphids for response to resistant genotypes of melon. 'Texas Resistant No. 1' melon was found to be resistant to one biotype of the melon aphid but susceptible to another.

Aphid resistance has been reported for cucumber, summer squash, watermelon, loofah and wild *Cucumis* species. Cucumber allele *bi* increases susceptibility to aphids.

The green peach aphid, *Myzus persicae*, also attacks cucurbits, but prefers other host plants. Green peach aphids frequently migrate from one cucurbit plant to another while seeking a more likable host, thereby spreading viruses.

The pumpkin aphid, *Illinoia cucurbita*, occurs on squash and melon in western USA.

Pickleworm

Pickleworm (*Diaphania nitidalis*) inhabits warm regions of southeastern USA, Mexico and the Caribbean islands. The larvae feed on blossoms and burrow into fruits of cucumber, melon and squash, reducing fruit set and yield and impairing fruit appearance. Secondary fruit rots invade damaged fruits. Similar damage is done by the melonworm, *D. hyalinata*, in southeastern USA. In India, *D. indica* is a pest on loofah.

Adult pickleworms and melonworms are about 2 cm long. The yellow-brown pickleworm moths and the silvery or white melonworm moths have a wing span of more than 2.5 cm. They fly at night and lay eggs on the lower surface of leaves, stems and fruits. Soon after hatching, pickleworm larvae tunnel into flowers and foliage and, later, burrow into fruits. Melonworm larvae generally feed on foliage. Young pickleworms are yellowish-white, and young melonworms are greenish-yellow. They grow to 16–22 mm long and hibernate as pupae, a cocoon encased in silk and often rolled up within a leaf. There may be several generations per year, with different stages of development present at the same time.

Resistance has been found in cucumber, melon and squash. Glabrous (*gl*), a cucumber allele that prevents formation of trichomes on foliage, has been associated with resistance to pickleworms. Evidently, the adult female relies on the trichomes of the cucumber plant as a stimulus for oviposition, and lays fewer eggs on glabrous plants. The melonworm has a similar reaction to glabrous melon plants. *Cucurbita moschata* cultivars are generally more resistant than those of *C. pepo*.

In some areas, pickleworms and melonworms can be avoided by early planting. Insecticides need to be applied early for good control, before the larvae tunnel into the protective plant tissue. *Bacillus thurigiensis* provides only moderate control. Deep ploughing of crop residues to kill pickleworm larvae is recommended. *Melothria pendula* and bitter melon, cucurbits which are weeds in Florida, are hosts for pickleworms and melonworms; their eradication can help control insect populations on nearby crops.

Melon Fruit Fly

Melon fruit fly (*Dacus cucurbitae*) is one of the most serious pests of cucurbits in India. It also occurs in Japan, southeastern Asia, East Africa and Hawaii, attacking melon, cucumber, watermelon, squash and other cucurbits.

Adult flies are yellow-brown and 8–10 mm long. The female fly oviposits into immature fruits. The larvae hatch within a week and feed on the fruit, causing damage with their tunnels and providing entry for fruit-rotting organisms. The larvae grow to a length of 10–12 mm, then the yellowish-white maggots drop to the ground and pupate in the soil.

Sometimes, developing fruits are covered as protection against fruit flies. For watermelon, dry grass mulching and paper wrapping are effective. Open paper bags are placed over bitter melon fruits in some countries.

Sources of resistance have been found in *Cucurbita maxima*, bottle gourd, both domesticated species of loofah, bitter melon, cucumber, melon and watermelon. Resistant melon fruits were found to have thicker, tougher rinds than melons susceptible to this pest. High silica content of the fruit rinds has also been associated with melon fruit fly resistance.

Oriental Fruit Fly

Oriental fruit fly (*Dacus dorsalis*) is closely related to the melon fruit fly. It is somewhat smaller, but otherwise looks like the melon fruit fly and has a similar life cycle. The oriental fruit fly has many more hosts than the melon fruit fly, including numerous fruit and vegetable crops in addition to cucurbits.

Oriental fruit fly is a serious pest in Hawaii and Asia. Stringent quarantine and eradication measures have been undertaken to prevent its introduction to California and Japan. Fumigation of fruits grown where the insect occurs is required before they can be shipped to quarantined areas. Baits that attract male flies are used to monitor the spread of the pest to new areas. When the flies are detected, eradication measures are quickly initiated.

Insects that parasitize the oriental fruit fly and sterilized male oriental fruit flies have been introduced for biological control. Other control measures include baits of sugar mixed with insecticide, removal of infested fruits and cultivation to disturb soilborne pupae.

Squash Bug

Squash bug (*Anasa tristis*) prefers plants of *Cucurbita*, but in the absence of a squash host, may feed on cucumber, melon, watermelon or other cucurbits. It is distributed only in the western hemisphere, from Canada to South America.

Feeding by squash bugs causes leaves of squash plants to turn yellow, then black. Severe infestations cause the plants to wilt, a condition known as 'Anasa wilt'. Nymphs and adults may damage or kill squash vines by sucking plant juices from leaves and fruits. Squash bugs have been reported to transmit the bacterial wilt pathogen.

Fig. 8.2. Squash bug nymphs (*Anasa tristis*) feeding on a cucurbit leaf. (Courtesy M.P. Hoffmann, photographer.)

Adults are winged, dark brown and about 1.75 cm long. Unmated adults overwinter, hibernating under dead vines and other protective debris. Large, yellow eggs are laid on the lower surface of leaves of young squash plants early the next season. The eggs turn brown, then hatch into pale-coloured nymphs. The nymphs differ from the adults by being soft-bodied, wingless and initially green, later becoming grey and developing wing pads (Fig. 8.2). The nymphs moult four times. There is usually one generation per year in temperate regions.

Rotation and field sanitation after growing a squash crop assist in controlling the squash bug. Adults and nymphs often hide under debris around the base of squash plants. Researchers have placed boards near the plants, as a means of monitoring the squash bug population and evaluating plants for resistance. Insecticides are more effective on young nymphs than on older nymphs and adults.

Cucurbita maxima is very susceptible to the squash bug, but *C. moschata* is resistant. Resistance has also been found in *C. argyrosperma*.

Wireworms

Wireworms live in the soil and feed on the roots of cucurbits and other plants. Several species occur, but Whitaker and Davis (1962) state that *Limonius canus* is the most serious pest. Damage is likely to be most severe with early plantings on light soils. Wireworms are the immature stages of click beetles. They are yellow to brown, measuring 25 mm long or less.

Thrips

Thrips, including the western flower thrip (*Frankliniella occidentalis*), the sugar beet thrip (*Heliothrips femoralis*) and other species, are destructive to melon, cucumber (especially in glasshouses) and other cucurbits. Although thrips are so small (1 mm long) as to be barely visible, they can occur in such large numbers that considerable damage is done. They frequently inhabit flowers, but also feed on foliage. Affected leaves have silvery areas with black, necrotic spots and margins that curl downward. Thrips are important vectors of tospoviruses. Weed management and insecticides are common control measures. Resistance to western flower thrips has been reported for cucumber.

Leafminer

Leafminer (*Liriomyza* spp.) larvae burrow beneath the epidermis, especially on the upper surface of cucurbit leaves. They leave a trail of damaged tissue behind that externally appears to be circuitous lines of lighter green colour than the rest of the leaf. The chlorotic depressions in the leaves become targets for fungal infection. Leafminers may also spread viruses.

The adult female, a small fly about 1.5 mm long, oviposits into the leaf tissue, generally at the tip or side of the leaf. After feeding inside the leaf, the larvae emerge to pupate on the leaf surface, on another plant part or on the ground.

Leafminers are difficult to manage with insecticides once they are insulated between the leaf surfaces. Biorational control methods that do not destroy the natural enemies of leafminers are preferred. Resistance to the vegetable leafminer, *Liriomyza sativae*, occurs in melon, summer squash and pumpkin cultivars.

Leafhoppers

Leafhoppers, including the melon leafhopper (*Empoasca abrupta*), the potato leafhopper (*E. fabae*) and several other species, attack cucurbits and many other plants. In western USA, the sugar beet leafhopper (*Circulifer tenellus*) may transmit curly top virus when it feeds on cucurbits, although it prefers other hosts. Leafhoppers also spread the phytoplasma that causes aster yellows disease.

Adults jump or fly when disturbed, hence the name leafhopper. Leafhopper adults and nymphs feed on the lower leaf surface by sucking sap from the phloem. The adults, which are wedge-shaped, green and about 34 mm long, overwinter in warm climates and migrate to other areas in the spring. Eggs are oviposited into veins or petioles. Nymphs are similar in appearance to adults, but lack wings.

Various insecticides are used against leafhoppers; repeated applications every 1–2 weeks are typically necessary. At night, light traps can catch large numbers of individuals.

Cucurbit Caterpillar

Cucurbit caterpillar (*Margaronia indica*) damages foliage and fruits of various cucurbits, including melon, in India, Thailand and other Asian countries. Fruits injured by this insect often rot prematurely. Resistance has been found in a primitive wild form of *Cucumis melo* previously known as *C. callosus*.

Darkling Ground Beetles

Darkling ground beetles (*Blapstinus* spp.) may attack melons, especially if the fruit is cracked, making the fruits unmarketable. The beetles also feed on the netting of undamaged melons, and occasionally will feed on young stems. The small (ca 1 cm), blackish beetles live in the soil or hide in debris on the soil surface.

Cutworms

Cutworms (species of *Agrotis* and other genera) feed on and damage young cucurbits. The caterpillars hide in the soil during the day, coming out at night to chew on plant stems. Adults, measuring 10 mm long, and the morphologically similar but smaller nymphs overwinter in field debris. After mating, eggs are oviposited into leaves. There may be several generations per year. Cutworms can be controlled by baits and soil applications of insecticides. No resistance has been reported for cucurbits.

Glasshouse Whitefly

Glasshouse whitefly (*Trialeurodes vaporariorum*), as its common name implies, is primarily a problem in the glasshouse, although it also inhabits cucurbit fields. Cucumber and other cucurbit plants can be stunted by glasshouse whiteflies sucking their plant juices. Heavily infested plants may become covered with a black, sooty mould growing on the whitefly excreta. This pest can transmit MYV to melon and cucumber.

Adult whiteflies reach 1.5 mm long and have pale yellow bodies and white wings. They prefer glasshouses and other warm, moist areas with suitable plants from their very wide range of hosts. Yellow-green eggs, measuring ca 0.5 mm long, are attached to the lower leaf surface by short stalks. The semi-transparent larvae, about 0.75 mm long, and the pupae also inhabit the lower leaf surface. Damage is done by sucking nymphs and adults.

Encarsia formosa is a predator of whiteflies and has been used to protect glasshouse tomato crops. The pubescence on cucumber leaves, however, deters *E. formosa* from efficiently parasitizing whiteflies on cucumber. Consequently, the cucumber glabrous allele (*gl*) has been proposed as a means of biological control for the glasshouse whitefly. Undesirable pleiotropic effects of *gl* have discouraged breeding cultivars homozygous for this allele, but heterozygous plants appear normal except for trichome density being reduced sufficiently to permit *E. formosa* to reduce whitefly populations.

Resistance to the glasshouse whitefly is known in wild *Cucumis* species, but these species are not compatible with cucumber or melon. There are no resistant cultivars.

Insecticidal soap sprays are used to contain this pest. Because whiteflies are attracted to the colour yellow, squares of yellow plastic coated with a sticky substance are placed in glasshouses as traps and to monitor population size to determine if insecticide application is needed.

Sweet Potato Whitefly

Sweet potato whitefly (*Bemisia tabaci*) transmits viruses, including SLCV and LIYV, to squash and melon. It also damages summer squash by causing the silver leaf disorder, which has become a serious problem in recent years in Florida, France and the Middle East. Affected plants have prominent leaf veins and a silvery appearance on the upper leaf surface. Fruit colour may become pale, and yield is often reduced. Silvering has been attributed to a toxic reaction to feeding by the nymphs. Summer squash cultivars differ in sensitivity to the toxic factor. Recently, an especially troublesome strain (poinsettia) of this insect has built up to enormous populations in the Imperial Valley of California, causing severe damage to melon and other crops despite all attempts to control it with pesticides. *Cucurbita lundelliana* is reportedly resistant to the sweet potato whitefly.

Leaf-footed Bug

Leaf-footed bug (*Leptoglossus australis*) infests cucumber and other cucurbits in Australia, Africa, India, southeastern Asia and China. Adults and nymphs puncture stems, flowers and young fruits. Feeding results in dark spots on the fruit, which may drop prematurely. Heavy feeding causes stem tips to die.

The adults are 20–25 mm long and dark brown to grey with a pale orange stripe. Eggs laid in rows on the stem hatch in about a week, and five nymphal instars develop during the next 30 days.

Repeated insecticide applications may be needed to achieve control. Sanitation, including disking or removing vines after harvest, reduces the number of insects overwintering. A pheromone that attracts the male leaf-footed bug is used to monitor population density.

Spider Mites

Spider mites (*Tetranychus urticae* and several related species) damage cucurbits by sucking plant juices. They cause distortion and chlorotic spotting of leaves and produce webbing on the foliage. Spider mites can reduce yield and impair fruit quality. They are often a major problem for glasshouse cucumbers, and can also be serious pests in cucurbit fields, especially during hot, dry weather. They occur nearly everywhere that cucurbits are grown, and can be windblown to distant areas.

Spider mites, although quite small, can be seen. Their presence is also detected by their webs and the pale green, stippled appearance of infested leaves. Eggs are laid in large numbers on the lower leaf surface. Numerous generations may be produced in a single year. Overwintering in the field is usually as adults.

Spider mites can migrate from grassy borders to cucurbit fields. Mowing and other disturbances that would stimulate this migration should be avoided. Acaricides are applied to cucurbits or fumigated in glasshouses to kill spider mites, but some spider mite populations may develop resistance to these chemicals. There are a number of insect predators of spider mites, and killing these beneficial insects with insecticides such as pyrethroids may result in a buildup of spider mite populations. Resistance to spider mites has been reported in cucumber, melon and watermelon.

APPENDIX: COMMON CUCURBIT NAMES AND THEIR SCIENTIFIC NAME EQUIVALENTS

SORTED BY COMMON NAME

Common name	Scientific equivalent(s)
Acorn squash	Particular cultivars of *Cucurbita pepo*
African horned cucumber	*Cucumis metuliferus*
Angled loofah	*Luffa acutangula*
Antidote vine	*Fevillea cordifolia*
Balsam apple	*Momordica balsamina*
Banana squash	Particular cultivars of *Cucurbita maxima*
Bell squash	Particular cultivars of *Cucurbita moschata*
Bitter melon	*Momordica charantia*
Bottle gourd	*Lagenaria siceraria*
Bryony	Various species of *Bryonia*
Buffalo gourd	*Cucurbita foetidissima*
Bur gherkin	*Cucumis anguria*
Cantaloupe	Particular cultivars of *Cucumis melo*
Casabanana	*Sicana odorifera*
Chayote	*Sechium edule*
Cheese pumpkin	Particular cultivars of *Cucurbita moschata*
Chinese snake gourd	*Trichosanthes kirilowii*
Citron	*Citrullus lanatus* var. *citroides*
Cochinchin gourd	*Momordica cochinchinensis*
Cocozelle	Particular cultivars of *Cucurbita pepo*
Colocynth	*Citrullus colocynthis*
Crookneck	Particular cultivars of *Cucurbita pepo, C. moschata*

Common name	Scientific equivalent(s)
Cucumber	*Cucumis sativus*
Cushaw	Particular cultivars of *Cucurbita argyrosperma*, *C. moschata*
Delicious squash	Particular cultivars of *Cucurbita maxima*
Egusi	*Citrullus lanatus* var. *citroides*; *C. colocynthis*; *Cucumeropsis mannii*
Fig leaf gourd	*Cucurbita ficifolia*
Fluted pumpkin	*Telfairia occidentalis*
Gherkin	*Cucumis anguria*; particular cultivars of *C. sativus*
He-zi-cao	*Actinostemma tenerum*
Hubbard	Particular cultivars of *Cucurbita maxima*
Indreni	*Trichosanthes lepiniana*
Ivy gourd	*Coccinia grandis*
Japanese snake gourd	*Trichosanthes ovigera*
Jiao-gu-lan	*Gynostemma pentaphyllum*
Kaksa	*Momordica dioica*
Lard plant	*Hodgsonia macrocarpa*
Lollipop climber	*Diplocyclos palmatus*
Loofah	Species of *Luffa*
Luo-guo-di	*Hemsleya amabilis*
Luo-han-guo	*Siraitia grosvenorii*
Malabar gourd	*Cucurbita ficifolia*
Marrow	Particular cultivars of *Cucurbita maxima*, *C. pepo*
Melon	*Cucumis melo*
Mi-mao-gua-lou	*Trichosanthes villosa*
Muskmelon	Particular cultivars of *Cucumis melo*
!nara	*Acanthosicyos horridus*
Ornamental gourd	Particular cultivars of *Cucurbita pepo*
Oyster nut	*Telfairia pedata*
Pickling melon	Particular cultivars of *Cucumis melo*
Pointed gourd	*Trichosanthes dioica*
Preserving melon	*Citrullus lanatus* var. *citroides*
Pseudo-fritillary	*Bolbostemma paniculatum*
Pumpkin	Particular cultivars of *Cucurbita pepo*, *C. maxima*, *C. moschata*, *C. argyrosperma*
Red hail stone	*Thladiantha dubia*
Round melon	*Praecitrullus fistulosus*
Scallop squash	Particular cultivars of *Cucurbita pepo*
Show pumpkin	Particular cultivars of *Cucurbita maxima*
Smooth loofah	*Luffa cylindrica*

Snake gourd	*Trichosanthes cucumerina*
Snake melon	Particular cultivars of *Cucumis melo*
Sponge plant	*Momordica angustisepala*
Squash	Species of *Cucurbita*, particularly *C. argyrosperma, C. maxima, C. moschata, C. pepo*
Squirting cucumber	*Ecballium elaterium*
Straightneck	Particular cultivars of *Cucurbita pepo*
Stuffing cucumber	*Cyclanthera pedata*
Summer squash	Particular cultivars of *Cucurbita pepo*
Teasel gourd	*Cucumis dipsaceus*
Tinda	*Praecitrullus fistulosus*
Turban squash	Particular cultivars of *Cucurbita maxima*
Vegetable marrow	Particular cultivars of *Cucurbita pepo*
Watermelon	*Citrullus lanatus*
Wax gourd	*Benincasa hispida*
White-seeded melon	*Cucumeropsis mannii*
Wild cucumber	*Echinocystis lobata*
Winter melon	Particular cultivars of *Cucumis melo*; *Benincasa hispida*
Winter squash	Particular cultivars of *Cucurbita argyrosperma, C. maxima, C. moschata, C. pepo*
Xishuangbanna gourd	*Cucumis sativus* var. *xishuangbannesis*
Zucchini	Particular cultivars of *Cucurbita pepo*

SORTED BY SCIENTIFIC NAME

Scientific name	Common name
Acanthosicyos horridus	!nara
Actinostemma tenerum	He-zi-cao
Benincasa hispida	Wax gourd, winter melon
Bolbostemma paniculatum	Pseudo-fritillary
Bryonia spp.	Bryony
Citrullus colocynthis	Colocynth, egusi
Citrullus lanatus	Watermelon
Citrullus lanatus var. *citroides*	Citron, egusi, preserving melon
Coccinia grandis	Ivy gourd
Cucumeropsis mannii	White-seeded melon, egusi
Cucumis anguria	Bur gherkin
Cucumis dipsaceus	Teasel gourd

Scientific name	Common name
Cucumis melo	Melon
Cucumis metuliferus	African horned cucumber
Cucumis sativus	Cucumber
Cucumis sativus var. *xishuangbannesis*	Xishuangbanna gourd
Cucurbita argyrosperma	Squash, pumpkin
Cucurbita ficifolia	Malabar gourd, fig leaf gourd
Cucurbita foetidissima	Buffalo gourd
Cucurbita maxima	Squash, pumpkin
Cucurbita moschata	Squash, pumpkin
Cucurbita pepo	Squash, pumpkin, gourd
Cyclanthera pedata	Stuffing cucumber
Diplocyclos palmatus	Lollipop climber
Ecballium elaterium	Squirting cucumber
Echinocystis lobata	Wild cucumber
Fevillea cordifolia	Antidote vine
Gynostemma pentaphyllum	Jiao-gu-lan
Hemsleya amabilis	Luo-guo-di
Hodgsonia macrocarpa	Lard plant
Lagenaria siceraria	Bottle gourd
Luffa acutangula	Angled loofah
Luffa cylindrica	Smooth loofah
Luffa spp.	Loofah
Momordica angustisepala	Sponge plant
Momordica balsamina	Balsam apple
Momordica charantia	Bitter melon
Momordica cochinchinensis	Cochinchin gourd
Momordica dioica	Kaksa
Praecitrullus fistulosus	Round melon, tinda
Sechium edule	Chayote
Sicana odorifera	Casabanana
Siraitia grosvenorii	Luo-han-guo
Telfairia occidentalis	Fluted pumpkin
Telfairia pedata	Oyster nut
Thladiantha dubia	Red hail stone
Trichosanthes cucumerina	Snake gourd
Trichosanthes dioica	Pointed gourd
Trichosanthes kirilowii	Chinese snake gourd
Trichosanthes lepiniana	Indreni
Trichosanthes ovigera	Japanese snake gourd
Trichosanthes villosa	Mi-mao-gua-lou

REFERENCES

Adams, P., Graves, C.J. and Winsor, G.W. (1992) Some responses of cucumber, grown in beds of peat, to N, K and Mg. *Journal of Horticultural Science* 67, 877–884.

Akoroda, M.O., Ogbechie-Odiaka, N.I., Adebayo, M.L., Ugwo, O.E. and Fuwa, B. (1990) Flowering, pollination and fruiting in fluted pumpkin (*Telfairia occidentalis*). *Scientia Horticulturae* 43, 197–206.

Anderson, A.P. (1894) The grand period of growth in fruit of *Cucurbita pepo* determined by weight. *Minnesota Botanical Series* 1, 238–279.

Andeweg, J.M. and Bruyn, J.W. de (1958) Breeding of non-bitter cucumbers. *Euphytica* 7, 13–20.

Andres, T.C. (1990) Biosystematics, theories on the origin, and breeding potential of *Cucurbita ficifolia*. In: Bates, D.M., Robinson, R.W. and Jeffrey, C. (eds) *Biology and Utilization of the Cucurbitaceae*. Cornell University Press, Ithaca, New York, pp. 102–119.

Arce-Ochoa, J.P., Dainello, F., Pike, L.M. and Drews, D. (1995) Field performance comparison of two transgenic summer squash hybrids to their parental hybrid line. *HortScience* 30, 492–493.

Aung, L.H., Ball, A. and Kushad, M. (1990) Developmental and nutritional aspects of chayote (*Sechium edule*, Cucurbitaceae). *Economic Botany* 44, 157–164.

Baird, J.R. and Thieret, J.W. (1988) The bur gherkin (*Cucumis anguria* var. *anguria*, Cucurbitaceae). *Economic Botany* 42, 447–451.

Benzioni, A., Mendlinger, S., Ventura, M. and Huyskens, S. (1993) Germination, fruit development, yield, and postharvest characteristics of *Cucumis metuliferus*. In: Janick, J. and Simon, J.E. (eds) *New Crops*. John Wiley and Sons, New York, pp. 553–557.

Bhave, M.R., Gupta, V.S. and Ranjekar, P.K. (1986) Arrangement and size distribution of repeat and single copy DNA sequences in four species of Cucurbitaceae. *Plant Systematics and Evolution* 152, 133–151.

Blancard, D., Lecoq, H. and Pitrat, M. (1994) *Colour Atlas of Cucurbit Diseases: Observation, Identification and Control*. John Wiley and Sons, New York, 299pp.

Bohn, G.W., Kishaba, A.N. and Toba, H.H. (1972) Mechanisms of resistance to melon aphid in a muskmelon line. *HortScience* 7, 281–282.

Cantliffe, D.J. and Phatak, S.C. (1975) Plant population studies with pickling cucumbers grown for once-over harvest. *Journal of the American Society for*

Horticultural Science 100, 464–466.

Cantliffe, D.J., Robinson, R.W. and Shannon, S. (1972) Promotion of cucumber fruit set and development by chlorflurenol. *HortScience* 7, 416–418.

Castetter, E.F. (1925) Horticultural groups of cucurbits. *Proceedings of the American Society for Horticultural Science* 22, 338–340.

Chakravarty, H.L. (1990) Cucurbits of India and their role in the development of vegetable crops. In: Bates, D.M., Robinson, R.W. and Jeffrey, C. (eds) *Biology and Utilization of the Cucurbitaceae.* Cornell University Press, Ithaca, New York. pp. 325–334.

Chee, P.P. and Slightom, J.L. (1991) Transfer and expression of cucumber mosaic virus coat protein gene in the genome of *Cucumis sativus. Journal of the American Society of Horticultural Science* 116, 1098–1102.

Coleman, R.G. (1995) Methods that help manage incidence of whitefly, virus. *Florida Grower and Rancher* 88(12), 41.

Cui, H. and Zhang, X. (1991) Cucumber cultivar improvement in the People's Republic of China. *Cucurbit Genetics Cooperative Report* 14, 5–7.

Cummings, M.B. and Jenkins, E.W. (1925) Hubbard squash in storage. *Bulletin of the Vermont Agricultural Experiment Station* 251, 1–35.

Darwin, C. (1906) *The Movements and Habits of Climbing Plants,* popular edn. John Murray, London, 208pp.

Davies, J.N. and Kempton, R.J. (1976) Some changes in the composition of the fruit of the glasshouse cucumber (*Cucumis sativus*) during growth, maturation and senescence. *Journal of the Science of Food and Agriculture* 27, 413–418.

Davis, J.M. (1994) Luffa sponge gourd production practices for temperate climates. *HortScience* 29, 263–266.

Deakin, J.R., Bohn, G.W. and Whitaker, T.W. (1971) Interspecific hyridization in *Cucumis. Economic Botany* 25, 195–211.

Decker, D.S. (1986) A biosystematic study of *Cucurbita pepo.* PhD thesis, Texas A&M University, College Station, Texas.

Decker, D.S. (1988) Origin(s), evolution, and systematics of *Cucurbita pepo* (Cucurbitaceae). *Economic Botany* 42, 4–15.

Decker-Walters, D.S., Walters, T.W., Cowan, C.W. and Smith, B.D. (1993) Isozymic characterization of wild populations of *Cucurbita pepo. Journal of Ethnobiology* 13, 55–72.

Delesalle, V.A. and Mooreside, P.D. (1995) Estimating the costs of allocation to male and female functions in a monoecious cucurbit, *Lagenaria siceraria. Oecologia* 102, 9–16.

Denna, D.W. (1963) The physiological genetics of the bush and vine habit in *Cucurbita pepo* L. squash. *Proceedings of the American Society for Horticultural Science* 83, 657–666.

De Winter, B. (1990) A new species of *Citrullus* (Benincaseae) from the Namib Desert, Namibia. *Bothalia* 20, 209–211.

Dittmer, H.J. and Talley, B.P. (1964) Gross morphology of tap roots of desert cucurbits. *Botanical Gazette* 125, 121–126.

Duke, J.A. and Ayensu, E.S. (1985) *Medicinal Plants of China, Volume 1.* Reference Publications, Algonac, Michigan, 362pp.

Ensminger, A.H., Ensminger, M.E., Konlande, J.E. and Robson, J.R.K. (1983) *Foods and Nutrition Encyclopedia, Volume 1.* Pegus Press, Clovis, California, 1208pp.

Etchells, J.L., Bell, T.A., Costilow, R.N., Hood, C.E. and Anderson, T.E. (1973) Influence of temperature and humidity on microbial, enzymatic, and physical changes of stored, pickling cucumbers. *Applied Microbiology* 26, 943–950.

Facciola, S. (1990) *Cornucopia: A Source Book of Edible Plants.* Kampong Publications, Vista, California, 667pp.

FAO (1995) *Production Yearbook for 1994.* No. 48. Food and Agriculture Organization of the United Nations, Rome.

Fellers, P.J. and Pflug, I.J. (1967) Storage of pickling cucumbers. *Food Technology* 21, 74–78.

Fisher, J.B. and Ewers, F.W. (1991) Structural responses to stem injury in vines. In: Putz, F.E. and Mooney, H.A. (eds) *The Biology of Vines.* Cambridge University Press, Cambridge, pp. 99–124.

Floris, E. and Alvarez, J.M. (1995) New sources for powdery mildew resistance in melon from Spanish local cultivars. *Cucurbit Genetics Cooperative Report* 18, 40–42.

Francis, F.J. and Thomson, C.L. (1965) Optimum storage conditions for Butternut squash. *Proceedings of the American Society for Horticultural Science* 86, 451–456.

Ganal, M. and Hemleben, V. (1986) Comparison of the ribosomal RNA genes in four closely related Cucurbitaceae. *Plant Systematics and Evolution* 154, 63–77.

Gathman, A.C. and Bemis, W.P. (1990) Domestication of buffalo gourd, *Cucurbita foetidissima.* In: Bates, D.M., Robinson, R.W. and Jeffrey, C. (eds) *Biology and Utilization of the Cucurbitaceae.* Cornell University Press, Ithaca, New York, pp. 335–348.

Goffinet, M.C. (1990) Comparative ontogeny of male and female flowers of *Cucumis sativus.* In: Bates, D.M., Robinson, R.W. and Jeffrey, C. (eds) *Biology and Utilization of the Cucurbitaceae.* Cornell University Press, Ithaca, New York, pp. 288–304.

Gonsalves, D., Chee, P., Provvidenti, R., Seem, R., and Slightom, J.L. (1992) Comparison of coat protein-mediated and genetically-derived resistance in cucumbers to infection by cucumber mosaic virus under field conditions with natural challenge inoculations by vectors. *Biotechnology* 10, 1562–1570

Gonsalves, C., Xue, B., Yepes, M., Fuchs, M., Ling, K., Namba, S., Chee, P., Slightom, J.L. and Gonsalves, D. (1994) Transferring cucumber mosaic virus-white leaf strain coat protein gene into *Cucumis melo* L. and evaluating transgenic plants for protection against infections. *Journal of the American Society for Horticultural Science* 119, 345–355.

Gounaris, I., Hardison, R.C. and Boyer, C.D. (1990) Restriction site and genetic map of *Cucurbita pepo* chloroplast DNA. *Current Genetics* 18, 273–275.

Graham, J.D. and Bemis, W.P. (1990) Interspecific trisomics of *Cucurbita moschata.* In: Bates, D.M., Robinson, R.W. and Jeffrey, C. (eds) *Biology and Utilization of the Cucurbitaceae.* Cornell University Press, Ithaca, New York, pp. 421–426.

Grimstad, S.O. and Frimanslund, E. (1993) Effect of different day and night temperature regimes on glasshouse cucumber young plant production, flower bud formation and early yield. *Scientia Horticulturae* 53, 191–204.

Hall, W.C. (1949) The effects of photoperiod and nitrogen supply on growth and reproduction in the gherkin. *Plant Physiology* 24, 753–769.

Hand, D.W. (1984) Crop responses to winter and summer CO_2 enrichment. *Acta Horticulturae* 162, 45–62.

Hanna, H.Y., Colyer, P.D., Kirkpatrick, T.L., Romaine, D.J. and Vernon, P.R. (1993)

Improving yield of cucumbers in nematode-infested soil by double-cropping with a resistant tomato cultivar, using transplants and nematicides. *Proceedings of the Florida State Horticultural Society* 106, 163–165.

Hayata, Y., Niimi, Y. and Iwasaki, N. (1995) Synthetic cytokinin – 1-(2-chloro-4-pyridyl)-3-phenylurea (CPPU) – promotes fruit set and induces parthenocarpy in watermelon. *Journal of the American Society for Horticultural Science* 120, 997–1000.

Heiser, C.B. Jr. (1979) *The Gourd Book*, paperback edn. University of Oklahoma Press, Norman, Oklahoma, 248pp.

Heiser, C.B. Jr. and Schilling, E.E. (1990) The genus *Luffa*: a problem in phytogeography. In: Bates, D.M., Robinson, R.W. and Jeffrey, C. (eds) *Biology and Utilization of the Cucurbitaceae*. Cornell University Press, Ithaca, New York, pp. 120–133.

Hochmuth, R.C. and Hochmuth, G.J. (1991) Nitrogen requirement for mulched slicing cucumbers. *Proceedings of the Soil and Crop Science Society of Florida* 50, 130–133.

Holroyd, R. (1914) Morphology and physiology of the axis in Cucurbitaceae. *Botanical Gazette* 78, 1–14.

Hopen, H.J. and Ries, S.K. (1962) The mutually compensating effect of carbon dioxide concentrations and light intensities on the growth of *Cucumis sativus* L. *Proceedings of the American Society for Horticultural Science* 81, 358–364.

Hopp, R.J., Merrow, S.B. and Elbert, E.M. (1960) Varietal differences and storage changes in β-carotene content of six varieties of winter squashes. *Proceedings of the American Society for Horticultural Science* 76, 568–576.

Huang, B., NeSmith, D.S., Bridges, D.C. and Johnson, J.W. (1995) Responses of squash to salinity, waterlogging, and subsequent drainage: II. Root and shoot growth. *Journal of Plant Nutrition* 18, 141–152.

Hughes, D.L. and Yamaguchi, M. (1983) Identification and distribution of some carbohydrates of the muskmelon plant. *HortScience* 18, 739–740.

Hutton, M.G. and Robinson, R.W. (1992) Gene list for *Cucurbita* spp. *Cucurbit Genetics Cooperative Report* 15, 102–109.

Jeffrey, C. (1967) Cucurbitaceae. In: Milne-Redhead, E. and Polhill, R.M. (eds) *Flora of Tropical East Africa, Volume 4*. Whitefriabs Press Ltd., London and Tonbridge, 157pp.

Jeffrey, C. (1990) Systematics of the Cucurbitaceae: an overview. In: Bates, D.M., Robinson, R.W. and Jeffrey, C. (eds) *Biology and Utilization of the Cucurbitaceae*. Cornell University Press, Ithaca, New York, pp. 3–9.

Jennings, P. and Saltveit, M.E. (1994) Temperature effects on imbibition and germination of cucumber (*Cucumis sativus*) seeds. *Journal of the American Society for Horticultural Science* 119, 464–467.

Kays, S.J. and Silva Dias, J.C. (1995) Common names of commercially cultivated vegetables of the world in 15 languages. *Economic Botany* 49, 115–152.

Kihara, H. (1951) Triploid water-melons. *Proceedings of the American Society of Horticultural Science.* 58, 217–230.

Kirkbride, J.H. Jr (1993) *Biosystematic Monograph of the Genus Cucumis (Cucurbitaceae)*. Parkway Publishers, Boone, North Carolina, 159pp.

Knavel, D.E. (1991) Productivity and growth of short-internode muskmelon plants at various spacings or densities. *Journal of the American Society for Horticultural Science* 116, 926–929.

Kramer, P.J. (1942) Species differences with respect to water absorption at low soil temperatures. *American Journal of Botany* 29, 828–832.

Krug, H. and Liebig, H.P. (1980) Diurnal thermoperiodism of the cucumber. *Acta Horticulturae* 118, 83–94.

Kumagai, M.H., Turpen, T.H., Weinzettl, N., della-Cioppa, G., Turpen, A.M., Donson, J., Hilf, M.E., Grantham, G.L., Dawson, W.O., Chow, T.P., Piatak, M. Jr. and Grill, L.K. (1993) Rapid, high-level expression of biologically active α-trichosanthin in transfected plants by an RNA viral vector. *Proceedings of the National Academy of Science USA* 90, 427–430.

Lee, J. (1994) Cultivation of grafted vegetables I. Current status, grafting methods, and benefits. *HortScience* 29, 235–239.

Liebig, H.P. (1980) A growth model to predict yield and economical figures of the cucumber crop. *Acta Horticulturae* 118, 165–174.

Lira-Saade, R. (1995) *Estudios Taxonómicos y Ecogeográficos de las Cucurbitaceae Latinoamericanas de Importancia Económica*. Systematic and Ecogeographic Studies on Crop Genepools. 9. International Plant Genetic Resources Institute, Rome, Italy, 281pp.

Lorenz, O.A. and Maynard, D.N. (1980) *Knott's Handbook for Vegetable Growers*. John Wiley and Sons, New York, 390pp.

Lower, R.L. and Nienhuis, J. (1990) Prospects for increasing yields of cucumbers via *Cucumis sativus* var. *hardwickii* germplasm. In: Bates, D.M., Robinson, R.W. and Jeffrey, C. (eds) *Biology and Utilization of the Cucurbitaceae*. Cornell University Press, Ithaca, New York, pp. 397–405.

Loy, J.B. and Broderick, C.E. (1990) Growth, assimilate partitioning, and productivity of bush and vine cultivars of *Cucurbita maxima*. In: Bates, D.M., Robinson, R.W. and Jeffrey, C. (eds) *Biology and Utilization of the Cucurbitaceae*. Cornell University Press, Ithaca, New York, pp. 436–447.

MacGillivray, J.H., Hanna, G.C. and Minges, P.A. (1942) Vitamin, protein, calcium, iron, and calorie yield of vegetables per acre and per acre man-hour. *Proceedings of the American Society for Horticultural Science* 41, 293–297.

McCollum, T.G., Cantliffe, D.J. and Parks, H.S. (1987) Flowering, fruit set, and fruit development in birdsnest-type muskmelons. *Journal of the American Society for Horticultural Science* 112, 161–164.

Mendlinger, S. (1994) Effect of increasing plant density and salinity on yield and fruit quality in muskmelon. *Scientia Horticulturae* 57, 41–49.

Merrick, L.C. (1990) Systematics and evolution of a domesticated squash, *Cucurbita argyrosperma*, and its wild and weedy relatives. In: Bates, D.M., Robinson, R.W. and Jeffrey, C. (eds) *Biology and Utilization of the Cucurbitaceae*. Cornell University Press, Ithaca, New York, pp. 77–95.

Merrick, L.C. and Bates, D.M. (1989) Classification and nomenclature of *Cucurbita argyrosperma*. *Baileya* 23, 94–102.

Metcalf, R.L. and Rhodes, A.M. (1990) Coevolution of the Cucurbitaceae and Luperini (Coleoptera: Chrysomelidae): basic and applied aspects. In: Bates, D.M., Robinson, R.W. and Jeffrey, C. (eds) *Biology and Utilization of the Cucurbitaceae*. Cornell University Press, Ithaca, New York, pp. 167–182.

Miller, C.H. and Wehner, T.C. (1989) Cucumbers. In: Eskin, N.A.M. (ed.) *Quality and Preservation of Vegetables*. CRC Press, Inc., Boca Raton, Florida, pp. 245–264.

Morton, J.F. (1967) The balsam pear – an edible, medicinal, and toxic plant. *Economic Botany* 21, 57–68.

Morton, J.F. (1971) The wax gourd, a year-round Florida vegetable with unusual keeping quality. *Proceedings of the Florida State Horticultural Society* 84, 104–109.

Munger, H. M. (1992) The significance of some traits and their combinations in the usage of U.S. cucumber varieties. *Cucurbit Genetics Cooperative Report* 15, 17–18.

Munger, H.M. and Robinson, R.W. (1991) Nomenclature of *Cucumis melo* L. *Cucurbit Genetics Cooperative Report* 14, 43–44.

Munger, H.M., Kyle, M.M. and Robinson, R.W. (1993). Cucurbits. In: *Traditional Crop Breeding Practices: an Historical Review to Serve as a Baseline for Assessing the Role of Modern Biotechnology.* Organisation for Economic Co-operation and Development, Paris, pp. 47–60.

Navazio, J.P. (1994) Utilization of high-carotene cucumber germplasm for genetic improvement of nutritional quality. PhD thesis, University of Wisconsin, Madison, USA.

Navot, N. and Zamir, D. (1986) Linkage relationships of 19 protein coding genes in watermelon. *Theoretical and Applied Genetics* 72, 274–278.

Navot, N. and Zamir, D. (1987) Isozyme and seed protein phylogeny of the genus *Citrullus* (Cucurbitaceae). *Plant Systematics and Evolution* 156, 61–67.

Nerson, H., Cantliffe, D.J., Paris, H.S. and Karchi, Z. (1982) Low-temperature germination of birdsnest-type muskmelons. *HortScience* 17, 639–640.

Newstrom, L.E. (1990) Origin and evolution of chayote, *Sechium edule*. In: Bates, D.M., Robinson, R.W. and Jeffrey, C. (eds) *Biology and Utilization of the Cucurbitaceae.* Cornell University Press, Ithaca, New York, pp. 141–149.

Newstrom, L.E. (1991) Evidence for the origin of chayote, *Sechium edule* (Cucurbitaceae). *Economic Botany* 45, 410–428.

Nijs, A.P.D. den and Visser, D. (1985) Relationships between African species of the genus *Cucumis* L. estimated by the production, vigour and fertility of F_1 hybrids. *Euphytica* 34, 279–290.

Nitsch, J.P., Kurtz, E.B. Jr., Liverman, J.L. and Went, F.W. (1952) The development of sex expression in cucurbit flowers. *American Journal of Botany* 39, 32–43.

Obiagwu, C.J. and Odiaka, N.I. (1995) Fertilizer schedule for yield of fresh fluted pumpkin (*Telfairia occidentalis*) grown in lower Benue river basin of Nigeria. *Indian Journal of Agricultural Sciences* 65, 98–101.

Orth, A.B., Teramura, A.H. and Sisler, H.D. (1990) Effects of ultraviolet-B on fungal disease development in *Cucumis sativus*. *American Journal of Botany* 77, 1188–1192.

Paris, H.S. (1986) A proposed subspecific classification for *Cucurbita pepo*. *Phytologia* 61, 133–138.

Paris, H.S. (1994) Genetic analysis and breeding of pumpkins and squash for high carotene content. In: Linskens, H-F. and Jackson, J.F. (eds) *Modern Methods of Plant Analysis, Vol. 16, Vegetables and Vegetable Products.* Springer-Verlag, Berlin, pp. 93–115.

Peet, M.M. and Willits, D.H. (1987) Glasshouse CO_2 enrichment alternatives: effects of increasing concentration or duration of enrichment on cucumber yields. *Journal of the American Society for Horticultural Science* 112, 236–241.

Perl-Treves, R. and Galun, E. (1985) The *Cucumis* plastome: physical map, intrageneric variation and phylogenetic relationships. *Theoretical and Applied Genetics* 71, 417–429.

Perry, K.B. and Wehner, T.C. (1990) Prediction of cucumber harvest date using a heat unit model. *HortScience* 25, 405–406.

Pier, J.W. and Doerge, T.A. (1995) Concurrent evaluation of agronomic, economic, and environmental aspects of trickle-irrigated watermelon production. *Journal of Environmental Quality* 24, 79–86.

Pierce, L.C. (1987) *Vegetables. Characteristics, Production, and Marketing.* John Wiley and Sons, New York, 433pp.

Pierce, L.K. and Wehner, T.C. (1990) Review of genes and linkage groups in cucumber. *HortScience* 25, 605–615.

Pitrat, M. (1994a) Gene list for *Cucumis melo* L. *Cucurbit Genetics Cooperative Report* 17, 135–147.

Pitrat, M. (1994b) Linkage groups in *Cucumis melo* L. *Cucurbit Genetics Cooperative Report* 17, 148.

Porterfield, W.M. (1955) Loofah – the sponge gourd. *Economic Botany* 9, 211–223.

Pratt, H.K., Goeschl, J.D. and Martin, F.W. (1977) Fruit growth and development, ripening, and the role of ethylene in the 'Honey Dew' muskmelon. *Journal of the American Society for Horticultural Science* 102, 203–210.

Provvidenti, R. (1990) Viral diseases and genetic sources of resistance in *Cucurbita* species. In: Bates, D.M., Robinson, R.W. and Jeffrey, C. (eds) *Biology and Utilization of the Cucurbitaceae.* Cornell University Press, Ithaca, New York, pp. 427–435.

Provvidenti, R. (1995) A multi-viral resistant cultivar of bottle gourd (*Lagenaria siceraria* from Taiwan). *Cucurbit Genetics Cooperative Report* 18, 65–67.

Provvidenti, R., Robinson, R.W. and Munger, H.M. (1978) Resistance in feral species to six viruses infecting *Cucurbita. Plant Disease Reporter* 62, 326–329.

Puchalski, J.T. and Robinson, R.W. (1990) Electrophoretic analysis of isozymes in *Cucurbita* and *Cucumis* and its application for phylogenetic studies. In: Bates, D.M., Robinson, R.W. and Jeffrey, C. (eds) *Biology and Utilization of the Cucurbitaceae.* Cornell University Press, Ithaca, New York, pp. 60–76.

Quemada, H.D. and Groff, D.W. (1995) Genetic engineering approaches in the breeding of virus resistant squash. In: Lester, G. and Dunlap, J. (eds), *Cucurbitaceae '94.* Gateway Printing, Edinburg, Texas, pp. 93–94.

Ramachandran, C. and Seshadri, V.S. (1986) Cytological analysis of the genome of cucumber (*Cucumis sativus* L.) and muskmelon (*Cucumis melo* L.). *Zeistschrift Pflanzenzuchtung* 96, 25–38.

Resh, H.M. (1987) *Hydroponic Food Production.* Woodbridge Press Publishing Company, Santa Barbara, California.

Reynolds, S.J. and Smith, S.M (1995) The isocitrate lyase gene of cucumber: isolation, characterization and expression in cotyledons following seed germination. *Plant Molecular Biology* 27, 487–497.

Rhodes, B. and Zhang, X. (1995) Gene list for watermelon. *Cucurbit Genetics Cooperative Report* 18, 69–84.

Robinson, R.W. (1992) Genetic resistance in the Cucurbitaceae to insects and spider mites. *Plant Breeding Reviews* 10, 309–360.

Robinson, R.W. and Shail, J.W. (1987) Genetic variability for compatibility of an interspecific cross. *Cucurbit Genetics Cooperative Report* 10, 88–89.

Robinson, R.W., Shannon, S. and Guardia, M.D. de la (1969) Regulation of sex expression in the cucumber. *BioScience* 19, 141–142.

Robinson, R.W., Whitaker, T.W. and Bohn, G.W. (1970) Promotion of pistillate flowering in *Cucurbita* by 2-chloroethylphosphonic acid. *Euphytica* 19, 180–183.

Robinson, R.W., Cantliffe, D.J. and Shannon, S. (1971) Morphactin-induced partheno-carpy in the cucumber. *Science* 171, 1251–1252.

Robinson, R.W., Whitaker, T.W., Munger, H.M. and Bohn, G.W. (1976) Genes of the cultivated Cucurbitaceae. *HortScience* 16, 554–568.

Rordan var Eysinga, J.P.N.L. and Smilde, K.W. (1969) *Nutritional Disorders in Cucumbers and Gherkins Under Glass*. Centre for Agricultural Publishing and Documentation, Wageningen, Holland, 46pp.

Roy, R.P. and Saran, S. (1990) Sex expression in the Cucurbitaceae. In: Bates, D.M., Robinson, R.W. and Jeffrey, C. (eds) *Biology and Utilization of the Cucurbitaceae*. Cornell University Press, Ithaca, New York, pp. 251–268.

Rudich, J. (1990) Biochemical aspects of hormonal regulation of sex expression in cucurbits. In: Bates, D.M., Robinson, R.W. and Jeffrey, C. (eds) *Biology and Utilization of the Cucurbitaceae*. Cornell University Press, Ithaca, New York, pp. 269–280.

Rundel, P.W. and Franklin, T. (1991) Vines in arid and semi-arid ecosystems. In: Putz, F.E. and Mooney, H.A. (eds) *The Biology of Vines*. Cambridge University Press, Cambridge, pp. 337–356.

Schales, F.D. and Isenberg, F.M. (1963) The effect of curing and storage on chemical composition and taste acceptability of winter squash. *Proceedings of the American Society for Horticultural Science* 83, 667–674.

Seshadri, V.S. (1986) Cucurbits. In: Bose, T.K. and Som, M.G. (eds) *Vegetable Crops in India*. Naya Prokash, Calcutta, India, pp. 91–164.

Shannon, S. and Robinson, R.W. (1979) The use of ethephon to regulate sex expression of summer squash for hybrid seed production. *Journal of the American Society of Horticultural Science* 104, 674–677.

Sharples, G.C. and Foster, R.E. (1958) The growth and composition of cantaloup plants in relation to the calcium saturation percentage and nitrogen level of the soil. *Proceedings of the American Society for Horticultural Science* 72, 417–425.

Sherman, M., Paris, H.S. and Allen, J.J. (1987) Storability of summer squash as affected by gene B and genetic background. *HortScience* 22, 920–922.

Shih, S. (1962) *A Preliminary Survey of the Book Ch'i Min Yao Shu: An Agricultural Encyclopaedia of the 6th Century*. 2nd edn. Science Press, Peking, China.

Singh, A.K. (1990) Cytogenetics and evolution in the Cucurbitaceae. In: Bates, D.M., Robinson, R.W. and Jeffrey, C. (eds) *Biology and Utilization of the Cucurbitaceae*. Cornell University Press, Ithaca, New York, pp. 10–28.

Singh, D. and Dathan, A.S.R. (1990) Seed coat anatomy of the Cucurbitaceae. In: Bates, D.M., Robinson, R.W. and Jeffrey, C. (eds) *Biology and Utilization of the Cucurbitaceae*. Cornell University Press, Ithaca, New York, pp. 225–238.

Singogo, W., Lamont, Jr., W.J. and Marr, C.W. (1996) Fall-planted cover crops support good yields of muskmelons. *HortScience* 31, 62–64.

Sinnott, E.W. (1932) Shape changes during fruit development in *Cucurbita* and their importance in the study of shape inheritance. *The American Naturalist* 66, 301–309.

Slack, G. and Hand, D.W. (1983) The effect of day and night temperatures on the growth, development and yield of glasshouse cucumbers. *Journal of Horticultural Science* 58, 567–573.

Soltani, N., Anderson, J.L. and Hamson, A.R. (1995) Growth analysis of watermelon plants grown with mulches and rowcovers. *Journal of the American Society for Horticultural Science* 120, 1001–1009.

Stocking, K.M. (1955) Some taxonomic and ecological considerations of the genus *Marah* (Cucurbitaceae). *Madroño* 13(4), 113–144.

Swiader, J.M., Sipp, S.K. and Brown, R.E. (1994) Pumpkin growth, flowering, and fruiting response to nitrogen and potassium sprinkler fertigation in sandy soil. *Journal of the American Society for Horticultural Science* 119, 414–419.

Tapley, W.T., Enzie, W.D. and van Eseltine, G.P. (1937) *The Vegetables of New York, Volume 1, Part IV, The Cucurbits*. New York State Agricultural Experiment Station, Geneva, 131pp.

Tindall, H.C. (1983) *Vegetables in the Tropics*. AVI Publishing Company, Westport, Connecticut, 533pp.

Trehane, P., Brickell, C.D., Baum, B.R., Hetterscheid, W.L.A., Leslie, A.C., McNeill, J., Spongberg, S.A. and Vrugtman, F. (eds) (1995) *International Code of Nomenclature for Cultivated Plants – 1995*. Quarterjack Publishing, Wimborne, United Kingdom, 175pp.

Wall, J.R. and York, T.L. (1960) Genetic diversity as an aid to interspecific hybridization in *Phaseolus* and in *Cucurbita*. *Proceedings of the American Society for Horticultural Science* 75, 419–428.

Walters, S.A. and Wehner, T.C. (1994) Evaluation of the U.S. cucumber germplasm collection for early flowering. *Plant Breeding* 112, 234–238.

Walters, S.A., Wehner, T.C. and Barker, K.R. (1993) Root-knot nematode resistance in cucumber and horned cucumber. *HortScience* 28, 151–154.

Walters, T.W. and Decker-Walters, D.S. (1988) Balsam-pear (*Momordica charantia*, Cucurbitaceae). *Economic Botany* 42, 286–288.

Walters, T.W. and Thieret, J.W. (1993) The snake melon (*Cucumis melo*; Cucurbitaceae). *Economic Botany* 47, 99–100.

Weeden, N.F. and Robinson, R.W. (1986) Allozyme segregation ratios in the interspecific cross *Cucurbita maxima* × *C. ecuadorensis* suggest that hybrid breakdown is not caused by minor alterations in chromosome structure. *Genetics* 114, 593–609.

Wehner, T.C. (1993) Gene list update for cucumber. *Cucurbit Genetics Cooperative Report* 16, 92–97.

Wehner, T.C. and Ellington, T.L. (1995) Growth regulator effects on sex expression of luffa sponge gourd. *Cucurbit Genetics Cooperative Report* 18, 68.

Wehner, T.C. and Humphries, E.G. (1995) A single-fruit seed extractor for cucumbers. *HortTechnology* 5, 268–273.

Whistler, W.A. (1990) The other Polynesian gourd. *Pacific Science* 44, 115–122.

Whitaker, T.W. and Davis, G.N. (1962) *Cucurbits. Botany, Cultivation, and Utilization*. Interscience Publishers, Inc., New York, 250pp.

Whitaker, T.W. and Robinson, R.W. (1986) Squash breeding. In: Bassett, M.J. (ed.) *Vegetable Breeding*. AVI Publishing Company, Westport, Connecticut, pp. 209–242.

Widders, I.E. and Price, H.C. (1989) Effects of plant density on growth and biomass partitioning in pickling cucumbers. *Journal of the American Society for Horticultural Science* 114, 751–755.

Wien, H.C. (1997) The cucurbits: cucumber, melon, squash and pumpkin. In: Wien,

H.C. (ed.) *The Physiology of Vegetable Crops*. CAB International, Wallingford, Oxon, England.

Wilson, H.D., Doebley, J. and Duvall, M. (1992) Chloroplast DNA diversity among wild and cultivated members of *Cucurbita* (Cucurbitaceae). *Theoretical and Applied Genetics* 84, 859–865.

Yang, R.Z. and Tang, C.S. (1988) Plants used for pest control in China: a literature review. *Economic Botany* 42, 376–406.

Yang, S.L. and Walters, T.W. (1992) Ethnobotany and the economic role of the Cucurbitaceae of China. *Economic Botany* 46, 349–367.

York, A. (1992) Pests of cucurbit crops: marrow, pumpkin, squash, melon and cucumber. In: McInlar, R.G.C. (ed.) *Vegetable Crop Pests*. CRC Press, Boca Raton, Florida, pp. 139–161.

Zhang, X. and Jiang, Y. (1990) Edible seed watermelons (*Citrullus lanatus* (Thunb.) Matsum. & Nakai) in Northwest China. *Cucurbit Genetics Cooperative Report* 13, 40–42.

Zitter, T.A., Hopkins, D.L. and Thomas, C.E. (eds) (1996) *Compendium of Cucurbit Diseases*. APS (American Phytopathological Society) Press, St. Paul, Minnesota, 87pp.

INDEX

Page numbers in **bold** are the primary sources of information for an entry.